Radiotherapy in Practice

Radiotherapy in Practice
Physics for Clinical Oncology

Edited by

Amen Sibtain
Department of Radiotherapy & Oncology,
St Bartholomew's Hospital,
West Smithfield,
London, UK

Andrew Morgan
Head of Radiotherapy Physics,
The Beacon Centre,
Musgrove Park Hospital,
Taunton, Somerset, UK

Niall MacDougall
Head of Clinical Dosimetry,
Radiotherapy Physics,
St Bartholomew's Hospital,
West Smithfield, London, UK

OXFORD
UNIVERSITY PRESS

OXFORD
UNIVERSITY PRESS

Great Clarendon Street, Oxford OX2 6DP
United Kingdom

Oxford University Press is a department of the University of Oxford.
It furthers the University's objective of excellence in research, scholarship,
and education by publishing worldwide. Oxford is a registered trade mark of
Oxford University Press in the UK and in certain other countries

First published 2012
Reprinted 2014

Published in the United States of America by Oxford University Press
198 Madison Avenue, New York, NY 10016, United States of America

British Library Cataloguing in Publication Data
Data available

Library of Congress Cataloging in Publication Data
Data available

ISBN 978-0-19-957335-6

Dedication

AS: For Isabella and Theo
NMD: For Emily
AMM: For Jean Morgan, Hilda Beswick, Marg Davies and more recently Kàren Morgan—who between them have managed to ensure that I've survived long enough to do something vaguely useful . . .

Foreword

Radiotherapy remains the most effective non-surgical treatment in the management of malignant disease. Over 50% of patients treated will have curative treatment and for the remainder effective palliation of pain, haemorrhage, obstructive symptoms and neurological complications can be achieved. In the years ahead, cancer will remain a major burden throughout the world with the incidence estimated to continue upward trends. In this setting, access to modern radiotherapy is an essential component of any healthcare system.

A sound knowledge of physics is an essential pre-requisite for those wishing to practice radiation oncology, yet it is often considered the most arduous component of training, particularly for clinicians. However, understanding the basic principles of radioactivity and the interaction of ionising radiation with matter lies at the heart of the therapeutic use of ionising radiation. The basis of modern radiation dosimetry is rooted in the fundamental principles of radiation physics and modern training programmes rightly retain this element in their requirements for aspiring radiation oncologists and radiographers as well as medical physicists.

Recent years have seen dramatic changes in the use of radiotherapy. The advent of intensity modulated radiotherapy, image guided radiotherapy, image guided brachytherapy and stereotactic radiotherapy each present new challenges and opportunities to those charged with their implementation. A firm grounding in the underlying physics concepts enables practitioners to embrace these developments and apply them to the clinical care of patients, who will benefit from ever increasing sophistication in the radiotherapy techniques available.

I am delighted to include this new title on radiation physics in the series Radiotherapy in Practice. It provides a comprehensive coverage of the subject addressing both the basic principles and their modern application in the clinical setting. It is unique in providing this information under a single cover and will, I am sure, be highly valued by trainees and practitioners in radiation oncology across the world.

Peter Hoskin
Mount Vernon Cancer Centre

Acknowledgements

We are grateful to Professor David Thwaites for providing the benefit of his experience, his advice and guidance on this project.

We are also particularly grateful to Nicola Wilson and Jenny Wright at OUP for their support, patience, advice and understanding.

Contents

Contributors

E Aird
Dept of Radiotherapy Physics,
Mount Vernon Hospital,
Rickmansworth Road,
Northwood, UK

P Bownes
Head of Brachytherapy Physics,
Medical Physics and Engineering,
St James's Institute of Oncology,
Bexley Wing, St James's University
Hospital,
Beckett Street, Leeds, UK

J Byrne
Department of Radiation Physics,
Northern Centre for Cancer Care,
Freeman Road, High Heaton,
Newcastle-upon-Tyne,
Tyne-and-Wear, UK

S Chittenden
Physics Department,
Royal Marsden NHS Foundation Trust,
Downs Road, Sutton,
Surrey, UK

S Colligan
Head of Radiotherapy Physics,
Cancer Services, Raigmore Hospital,
Inverness, UK

G Flux
Physics Department,
Royal Marsden NHS Foundation Trust,
Downs Road, Sutton,
Surrey, UK

T Greener
Department of Radiotherapy Physics,
St Thomas Hospital,
Westminister Bridge Road,
London, UK

A Hounsell
Department of Radiotherapy Physics,
Northern Ireland Cancer Centre,
Belfast City Hospital, Belfast, UK

T Jordan
Head of Radiotherapy Physics,
St. Luke's Cancer Centre, Royal Surrey
County Hospital, Guildford,
Surrey, UK

V Khoo
Royal Marsden Hospital,
Fulham Road, London, UK

C Lee
Medical Physics Department,
Clatterbridge Centre for Oncology,
Clatterbridge Road, Bebington,
Wirral, Merseyside, UK

N MacDougall
Head of Clinical Dosimetry,
Radiotherapy Physics,
St Bartholomew's Hospital,
West Smithfield, London, UK

R Mackay
Department of Radiotherapy Physics,
The Christie Hospital,
Wilmslow road, Manchester, UK

J Mills
Radiotherapy Physics Manager,
University Hospital, Coventry, UK

A Morgan
Head of Radiotherapy Physics,
The Beacon Centre, Musgrove Park
Hospital, Taunton, Somerset, UK

C Nalder
Joint Department of Physics,
Royal Marsden Hospital, London, UK

A Nisbet
Head of Medical Physics,
Royal Surrey County Hospital,
Egerton Road, Guildford, UK

G Pitchford
Department of Radiotherapy Physics,
Lincoln County Hospital,
Greetwell Road, Lincoln,
Lincolnshire, UK

B Pratt
Physics Department,
Royal Marsden NHS Foundation Trust,
Downs Road, Sutton, Surrey, UK

C Richardson
Principal Physicist, Brachytherapy,
Bexley Wing, St James's University
Hospital, Alma Street, Leeds, UK

A Sibtain
Department of Radiotherapy &
Oncology, St Bartholomew's Hospital,
West Smithfield, London, UK

C Taylor
Department of Medical Physics,
Wellcome Wing, Leeds General
Infirmary, Great George Street, Leeds,
West Yorkshire, UK

N Van As
The Royal Marsden Hospital,
Fulham Road, London, UK

M Waller
Department of Medical Physics,
Wellcome Wing, Leeds General
Infirmary, Great George Street,
Leeds, West Yorkshire, UK

G Workman
Brachytherapy Lead,
Radiotherapy Physics Service,
Northern Ireland Cancer Centre,
Belfast City Hospital, Belfast, UK

Chapter 1

Introduction

A Sibtain, A Morgan and N MacDougall

Everything should be made as simple as possible, but no simpler.
Quotation attributed to Albert Einstein (1933)

Radiotherapy physics is often regarded as challenging and intimidating for the clinical oncologist, given that it is a subject that has not been studied for many years before entering the speciality. This is accentuated by the rapidly increasing complexity in modern radiotherapy; the great improvements in recent years are a direct result of the considerable escalation in computer processing power and improvements in precision engineering. The major new techniques have only been realized through this advance in technology.

This book aims to take the clinical oncologist in training though the essentials of radiotherapy physics without the need for large, inaccessible texts. Whilst primarily directed at clinicians, it will also be useful for trainee physicists, dosimetrists and therapy radiographers. It covers important, clinically relevant issues that underpin the principles of radiation therapy and the functioning of a modern radiotherapy department.

The structure of the book is generally, but not exclusively, based on the Royal College of Radiologists (UK) physics syllabus, which has benchmarked the important aspects of physics for clinical oncologists in the UK and beyond.

The first three chapters cover the essentials of basic physics, photon production and interaction, and the physics of charged and uncharged particles relevant to therapy. An innovative chapter, IT in Radiotherapy, demystifies current computer technology relevant to the radiotherapy department for the oncologist.

The principles of imaging for radiotherapy are summarized, followed by more detailed consideration of X-ray and electron beam physics. A guide to radiotherapy treatment planning, describing the processes, principles and methods of generating a radiotherapy treatment plan provides a comprehensive guide to simple and complex techniques for the clinician.

Beam therapy equipment, its design, rational and capabilities are described in Chapter 11, followed by the physics of brachytherapy, and sealed and unsealed sources.

As well as treatment delivery, radiation protection and quality assurance are vital for the oncologist to know and understand, providing insight into the invaluable work done in radiotherapy departments to ensure safety to staff and patients.

To achieve concision, we have deliberately minimized references other than established multidisciplinary and legal documents. Further information on any aspects should be readily available though other sources. We have also worked to minimize overlap with other books in the Radiotherapy in Practice series, and referred the reader to these other volumes where necessary.

The authors have been chosen for both their recognized expertise and for their ability to communicate complex concepts clearly, and we are grateful for their skilled and dedicated contributions. We hope the book fills a gap for the clinical oncologist as they embark on learning and understanding the raw principles and practice of radiation therapy delivery.

Chapter 2

Basic physics essentials for the radiation oncologist

A Sibtain, A Morgan and N MacDougall

2.1 The absolute basics

2.1.1 Elements and compounds

Everything is made up of matter. There are two types of matter—elements and compounds.

2.1.1.1 Traditional definition of an element

An element is a kind of matter that **cannot** be decomposed into two or more simpler types of matter. An example of an element is hydrogen.

2.1.1.2 Definition of a compound

A compound is a kind of matter that can be decomposed into two or more simpler types of matter.

2.1.1.3 An alternative definition of a compound

A compound is formed when two or more elements combine to produce a more complex kind of matter. An example of a compound is water, which can be broken down into the two elements, hydrogen and oxygen (Fig 2.1).

2.1.2 Atoms and molecules

Atoms are the very smallest particles of an element that can exist without losing the chemical properties of the element. There are 114 types of atom, all defined in the periodic table (Fig. 2.2) by their atomic numbers. The periodic table arranges the atoms in groups and in periods. The rows are called periods and the columns are called groups. Elements/atoms in the same groups are similar to each other.

Molecules are the smallest particles of a compound that can exist without losing the chemical properties of that compound—for example, the water molecule consisting of two hydrogen atoms and one oxygen atom. If the molecule is broken down further the resulting matter loses the properties of water.

2.1.2.1 Atomic substructure

Atoms can be broken down into smaller particles. These particles are neutrons, protons and electrons. Neutrons and protons are in the nucleus of the atom and are surrounded

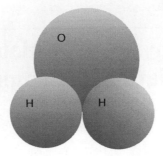

Fig. 2.1 Water Molecule. Water is a compound, the smallest part of which is the water molecule. If it is broken down further into hydrogen and oxygen atoms, it loses the properties of water.

by the electrons. Protons are relatively large particles and have a positive charge. Neutrons are also 'large' but have no charge. Electrons are relatively much smaller and lighter particles. They are attracted to the nucleus because they have a negative charge, but do not collide with it because the electrons orbit the nucleus (Table 2.1).

It is easy to imagine the atom as a group of billiard balls in the nucleus and a smaller ball—the electron—moving around the nucleus in the same way the planets of the solar system orbit the sun—so a hydrogen atom would look as shown in Fig 2.3.

There are two things wrong with this way of imagining atoms. First, the electron is actually much further away from the nucleus than the figure depicts—if it were drawn to scale it would be about 500 meters away. An atom is therefore mostly space. Second, whilst protons and neutrons do behave as particles, electrons also have wave type behaviour, like light. Rather than moving in a fixed path around the nucleus like a 'planet', there is a cloud of electron waves around the nucleus. The wave/particle paradox

hydrogen 1 **H** 1.0079												boron 5 **B** 10.811	carbon 6 **C** 12.011	nitrogen 7 **N** 14.007	oxygen 8 **O** 15.999	fluorine 9 **F** 18.998	helium 2 **He** 4.0026 neon 10 **Ne** 20.180
lithium 3 **Li** 6.941	beryllium 4 **Be** 9.0122											aluminium 13 **Al** 26.982	silicon 14 **Si** 28.086	phosphorus 15 **P** 30.974	sulfur 16 **S** 32.065	chlorine 17 **Cl** 35.453	argon 18 **Ar** 39.948
sodium 11 **Na** 22.990	magnesium 12 **Mg** 24.305	scandium 21 **Sc** 44.956	titanium 22 **Ti** 47.867	vanadium 23 **V** 50.942	chromium 24 **Cr** 51.996	manganese 25 **Mn** 54.938	iron 26 **Fe** 55.845	cobalt 27 **Co** 58.933	nickel 28 **Ni** 58.693	copper 29 **Cu** 63.546	zinc 30 **Zn** 65.39	gallium 31 **Ga** 69.723	germanium 32 **Ge** 72.61	arsenic 33 **As** 74.922	selenium 34 **Se** 78.96	bromine 35 **Br** 79.904	krypton 36 **Kr** 83.80
potassium 19 **K** 39.098	calcium 20 **Ca** 40.078	yttrium 39 **Y** 88.906	zirconium 40 **Zr** 91.224	niobium 41 **Nb** 92.906	molybdenum 42 **Mo** 95.94	technetium 43 **Tc** [98]	ruthenium 44 **Ru** 101.07	rhodium 45 **Rh** 102.91	palladium 46 **Pd** 106.42	silver 47 **Ag** 107.87	cadmium 48 **Cd** 112.41	indium 49 **In** 114.82	tin 50 **Sn** 118.71	antimony 51 **Sb** 121.76	tellurium 52 **Te** 127.60	iodine 53 **I** 126.90	xenon 54 **Xe** 131.29
rubidium 37 **Rb** 85.468	strontium 38 **Sr** 87.62	lutetium 71 **Lu** 174.97	hafnium 72 **Hf** 178.49	tantalum 73 **Ta** 180.95	tungsten 74 **W** 183.84	rhenium 75 **Re** 186.21	osmium 76 **Os** 190.23	iridium 77 **Ir** 192.22	platinum 78 **Pt** 195.08	gold 79 **Au** 196.97	mercury 80 **Hg** 200.59	thallium 81 **Tl** 204.38	lead 82 **Pb** 207.2	bismuth 83 **Bi** 208.98	polonium 84 **Po** [209]	astatine 85 **At** [210]	radon 86 **Rn** [222]
caesium 55 **Cs** 132.91	barium 56 **Ba** 137.33	57-70 ✱															
francium 87 **Fr** [223]	radium 88 **Ra** [226]	89-102 ✱✱	lawrencium 103 **Lr** [262]	rutherfordium 104 **Rf** [261]	dubnium 105 **Db** [262]	seaborgium 106 **Sg** [266]	bohrium 107 **Bh** [264]	hassium 108 **Hs** [269]	meitnerium 109 **Mt** [268]	ununnilium 110 **Uun** [271]	unununium 111 **Uuu** [272]	ununbium 112 **Uub** [277]	ununquadium 114 **Uuq** [289]				

	lanthanum 57 **La** 138.91	cerium 58 **Ce** 140.12	praseodymium 59 **Pr** 140.91	neodymium 60 **Nd** 144.24	promethium 61 **Pm** [145]	samarium 62 **Sm** 150.36	europium 63 **Eu** 151.96	gadolinium 64 **Gd** 157.25	terbium 65 **Tb** 158.93	dysprosium 66 **Dy** 162.50	holmium 67 **Ho** 164.93	erbium 68 **Er** 167.26	thulium 69 **Tm** 168.93	ytterbium 70 **Yb** 173.04
✱Lanthanide series														
✱✱ Actinide series	actinium 89 **Ac** [227]	thorium 90 **Th** 232.04	protactinium 91 **Pa** 231.04	uranium 92 **U** 238.03	neptunium 93 **Np** [237]	plutonium 94 **Pu** [244]	americium 95 **Am** [243]	curium 96 **Cm** [247]	berkelium 97 **Bk** [247]	californium 98 **Cf** [251]	einsteinium 99 **Es** [252]	fermium 100 **Fm** [257]	mendelevium 101 **Md** [258]	nobelium 102 **No** [259]

Fig. 2.2 The periodic table.

Table 2.1 The charge and mass of sub-atomic particles

Name	Charge	Mass
Proton	Positive	$1.6726*10^{-27}$ kg
Neutron	Neutral	$1.6929*10^{-27}$ kg
Electron	Negative	$9.11*10^{-31}$ kg

has been thoroughly explored over the past century. However, for the purposes of radiotherapy, electrons should be imagined as particles.

2.1.2.2 Atomic and mass numbers

Each atom has a particular number of protons and neutrons.

◆ The atomic number is the number of protons in the nucleus, Z.

◆ The mass number of an atom is the number of protons and neutrons added together, A.

The atomic and mass numbers for an atom X are depicted as:

$$_{Z}^{A} X$$

The atomic number, i.e. the number of protons in an atom, defines the atom/element. If the number of protons is somehow changed, the atom changes into that of another element.

In contrast, if the number of neutrons is changed, the atom remains the same, but may have some different characteristics. Atoms with the same atomic number but different mass numbers are called **isotopes**.

> The total number of protons and neutrons determine the nuclide. The number of neutrons relative to the protons determines the stability of the nucleus, with certain isotopes undergoing radioactive decay.

2.1.3 Electron shells and energy levels

Electrons reside around the nucleus in a number of 'shells'. They cannot exist between these shells.

The shells are labelled with letters of the alphabet, starting with K at the inner shell (Fig. 2.4).

Each shell can hold a maximum number of electrons (Table 2.2).

Most shells are made up of sub-shells.

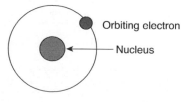

Fig. 2.3 A hydrogen atom.

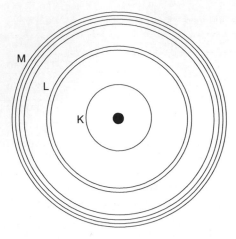

Fig. 2.4 Diagram of shells.

The shell closest to the nucleus (K) has one shell, which can hold a maximum of 2 electrons. The next shell out (L) has two sub-shells—one holding a maximum of 2 and the second capable of holding a maximum of 6 electrons. The next shell (M) has 3 sub-shells, holding 2, 6 and 10 electrons respectively. This goes on, and is summarized in the Table 2.3.

2.1.3.1 Binding energy

The electrons are held in their shell by their electrostatic attraction to the positively charged nucleus. To remove an electron from a shell, a certain amount of energy is needed to overcome this attraction. This is called the **binding energy.** The binding energy is greatest for the inner shell and is progressively lower for each shell moving away from the nucleus (Fig 2.5). Binding energies are greater for atoms with a greater number of protons in the nucleus (i.e. a higher atomic number) because they have a higher positive nuclear charge, and therefore a greater hold on the orbiting electrons. If an electron gains more energy than the binding energy, it can escape from the attraction of the nucleus and leave the atom. This is called **ionization**. The resulting

Table 2.2 Table of electron shell labels and the number of elecrons in each

Shell	Number of electrons
K	2
L	8
M	18
N	32
O	32
P	32
Q	32

Table 2.3 Sub-shells and their electron capacity

Sub-shell	Maximum electron capacity	Shells containing it
s	2	K
p	6	K, L
d	10	K, L, M
F	14	K, L, M, N
S	2	K
G	18	K, L, M, N, O

atom has a net positive charge because it has one less electron than it has protons— i.e. it is a positive ion.

2.1.3.2 Energy levels

An electron can also move between shells of different binding energies. This happens when an electron gains enough energy to move from one (sub-) shell to another, but not quite enough to escape the atom completely. Each (sub-) shell can therefore be seen as a fixed energy level and electrons can only exist in these shells if they possess that particular amount of energy. The energy levels are fixed for any particular type of atom.

As well as moving from a lower energy level to a higher energy level by gaining energy from somewhere, electrons can move the other way and release their excess energy (Fig. 2.6). 1 electron volt (eV) is equal to 1.602 x10^{-19} Joules.

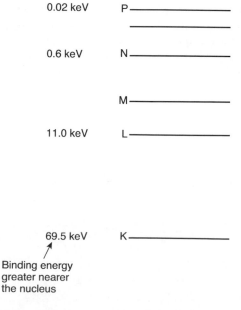

Fig. 2.5 Diagram of binding energy levels for a Tungsten atom.

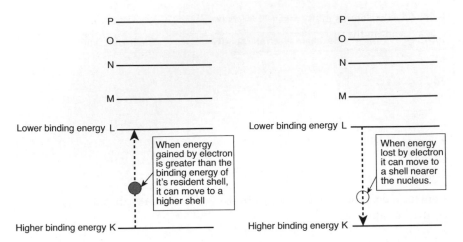

Fig. 2.6 Diagram of electron energy level movement.

2.1.3.3 **Energy levels for isolated atoms and interacting atoms**

The energy levels are exact for isolated atoms. However, when a number of atoms are together, such as in a molecule or as a solid element, the electron shells interact. This interaction allows each fixed energy level to expand to a range, which is called an energy band (Fig. 2.7).

2.1.4 **Key points for the radiotherapist**

- The movement of electrons between energy bands is the basis of X-ray production.
- Electrons, protons, and neutrons are used in cancer therapy. Their behaviour in tissue is related, in part, to their mass.
- Atoms of the same element have the same number of protons.

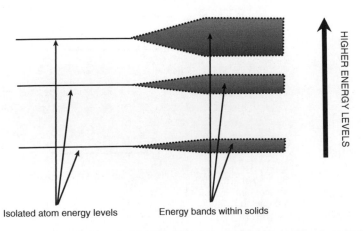

Fig. 2.7 Diagram of energy bands within solids.

♦ Nuclides are atoms of the same element with varying numbers of neutrons. This is relevant in the understanding of radionuclides because the relative number of protons to neutrons determines how stable the nucleus is.

♦ Ionization is an important process in the interaction of radiation with biological molecules.

2.2 Electromagnetism, electromagnetic radiation and the electromagnetic spectrum

There are four fundamental forces of nature:

♦ Gravity,

♦ Electromagnetism,

♦ Weak interaction,

♦ Strong interaction.

They are termed 'fundamental' because, they cannot be explained or picked apart by other forces. A number of theories exist to understand and unify these fundamental forces, but these are beyond the scope of this book and not needed to understand radiotherapy physics.

2.2.1 Electromagnetism

Electromagnetism is one of these fundamental forces and it describes the force of a magnetic field on moving charged particle. Similarly, it describes how moving charged particles create magnetic fields.

2.2.2 Electromagnetic radiation

Electromagnetic radiation is a form of energy transfer though space as a combination of electrical and magnetic fields. A moving electrical field generates a varying magnetic field and vice versa. These combined moving fields form the electromagnetic wave.

The inexplicable feature of electromagnetic radiation is that it sometimes behaves as waves and sometimes behaves as particles—summed up in the term 'wave-particle duality'.

2.2.3 The wave model of electromagnetic radiation

Electromagnetic radiation causes effects that suggest it behaves as waves. For example, it exhibits reflection, refraction and interference. All electromagnetic waves travel at a velocity of 3×10^8 metres per second in a vacuum.

2.2.4 Waves

Waves (Fig. 2.8) are a series of peaks and troughs and have definable features:

♦ Wavelength,

♦ Frequency,

♦ Energy.

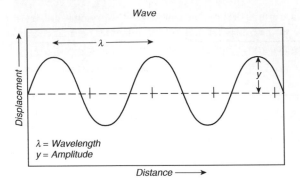

Fig. 2.8 Diagram of a wave.

Wavelength is the distance between two successive crests or troughs. The symbol is λ and it is measured in metres.

Frequency is the number of waves passing a particular point in unit time. The symbol is v and the unit is number per second or hertz (Hz).

The amplitude can be thought of as the energy of the wave.

2.2.5 **The particle behaviour of electromagnetic radiation**

Electromagnetic radiation also behaves as particles. These particles are discrete packets of energy and are called photons. The energy of these photons is proportional to the frequency of the electromagnetic wave to which they are linked. There is an equation that relates the energy and frequency—the Planck-Einstein equation,

$$E = h.v$$

where E is energy, h is Planck's constant (6.626×10^{-34} Joules per second ($J\ s^{-1}$)) and v is frequency so if frequency is the velocity divided by the wavelength,

$$E = h.c/\lambda$$

where c is the speed of light and λ is the wavelength of the wave.

So a short wavelength relates to high energy photons and a long wavelength to low energy photons.

In the realm of electromagnetic radiation, the velocity is constant so frequency and wavelength vary together. At high frequencies and short wavelengths, and therefore higher energies, electromagnetic radiation has more particle-like behaviour.

The range of frequency and wavelengths is called the electromagnetic spectrum (Fig. 2.9). Humans have evolved to detect part of this spectrum—visible light. The rest of the electromagnetic spectrum on either side of either side of visible light cannot be sensed.

2.3 **Radioactivity**

2.3.1 **Introduction to atomic structure**

Atoms consist of a central nucleus surrounded by an electron cloud. The nucleus is made up of neutrons and protons, held together by a strong force, called the 'strong

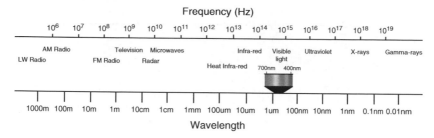

Fig. 2.9 The electromagnetic spectrum. This figure is reproduced in colour in the colour plate section.

nuclear force' to make it easy to remember! The 'strong nuclear force' is one of the fundamental forces of nature. The weak nuclear force also exists to hold sub-atomic particles together, and the electrostatic force is involved in holding the electron cloud around the nucleus.

2.3.2 The essence of radioactivity

The sub-atomic particles exist in a particular arrangement. The amount of energy in the particles can vary with the arrangement. They will always try to settle in an arrangement that has the lowest energy configuration. Some nuclides have unstable nuclear arrangements and shift to a more stable arrangement over time. While undergoing this rearrangement they emit one of the following:

♦ An alpha particle: consisting of two protons and two neutrons.

♦ A beta particle: an electron.

♦ A gamma ray: a packet of electromagnetic energy i.e. a photon.

Any element that undergoes this process is called radioactive, and the phenomenon is called radioactivity.

Another way of looking at radioactive materials is that they continuously emit energy in the form of the alpha particles, beta particles or electromagnetic waves.

Summary Radioactivity is the spontaneous decay of the nucleus of an atom from which either alpha, beta or gamma rays are emitted, though all processes may be occurring simultaneously in a sample of radioactive material.

2.3.3 The decay series: parent and daughter

Radioactive materials undergo a series of transformations until they reach their stable state.

These transformations occur in a series of steps. These steps are called a 'decay series'. The original element is called the 'parent' and the stable 'end-result' element is the 'stable daughter'. Any isotopes or elements in between are the 'excited daughter'.

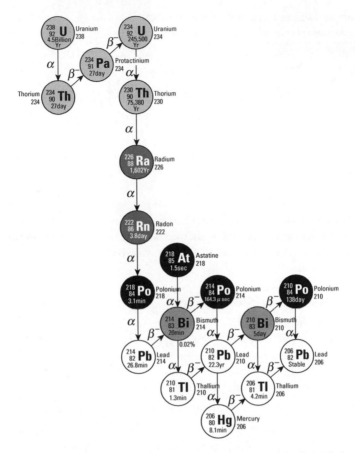

Fig. 2.10 The uranium decay series, U238 being the 'parent' and lead 206 being the stable daughter. All other elements in the series are the 'excited daughters'.

2.3.3.1 Alpha decay

Alpha decay is the emission of alpha particles by a nucleus.

Alpha particles are made up of 2 neutrons and 2 protons (the same as a Helium nucleus). They have a charge of +2 because each proton has a positive charge and neutrons have no charge.

They are slow moving and relatively heavy particles. They can be deflected in a magnetic field and can be stopped easily by a sheet of paper. They can cause large amounts of ionization inside the body.

Only heavy nuclei are able to undergo alpha decay. It results in the formation of a new element because of the loss of protons.

When an atom loses an alpha particle the atomic number reduces by 2 (i.e. the number of protons lost) and the mass number reduces by 4 (i.e. the number of protons plus the number of neutrons that are lost).

For example, radium has a mass number of 226 and an atomic number of 88. It undergoes alpha decay, losing 2 protons and 2 neutrons and becomes radon, which has a mass number of 222 and an atomic number of 86.

$$^{226}_{88}\text{Ra} \rightarrow\ ^{222}_{86}\text{Rd} + \alpha^{2+}$$

2.3.3.2 Beta decay

Beta particles can carry either a positive charge or a negative charge. Those with a negative charge are electrons and those with a positive charge are called positrons (the anti-particle of the electron).

These particles originate from the nucleus. They can travel for a few meters in air but are easily stopped by a sheet of aluminum or glass.

The emission of a negative beta particle, i.e. an electron, is called beta minus decay and the emission of a positive beta particle is called beta positive decay.

2.3.3.3 Beta minus decay

Beta minus decay is when a neutron in the nucleus converts into a proton and an electron (e⁻).

The electron is emitted (along with another particle called an anti-neutrino).

This conversion means the mass number stays the same i.e. the loss of the neutron is offset by the gain of a proton. However, the atomic number increases by 1, i.e. the net gain of one proton.

The beta particle is released with a certain amount of kinetic energy. The maximum possible kinetic energy is equal to the difference in the mass between the original nucleus and the post emission nucleus. Not all beta particles carry the maximum possible amount of kinetic energy—it is usually less than that. The remaining energy released is carried by the anti-neutrino.

$$^{137}_{55}\text{Cs} \rightarrow\ ^{137}_{56}\text{Ba} + e^- + \text{neutrino}$$

2.3.3.4 Beta positive decay

Beta positive decay is when a proton in the nucleus is converted into a neutron and a positron (e⁺). The positron is emitted along with another particle called a neutrino. Neutrinos and anti-neutrinos are of great interest to particle physicists but are of no relevance in radiotherapy. The atomic number reduces by one due to a net loss of protons, but the mass number remains the same because the total number of protons and neutrons does not change.

Positrons, once released, lose their kinetic energy. When most of this has gone, the positron combines with an electron. Their combined rest mass turns into two 511 keV photons each travelling in opposite directions from the point of the annihilation.

This is the key process underlying Positron Emission Tomography (PET) imaging.

$$_{9}^{18}F \rightarrow _{8}^{18}O + e^+ + \text{neutrino}$$

2.3.4 Electron capture

In electron capture, the nucleus combines with one of the orbiting electrons, converting one of its protons into a neutron, and releasing a neutrino. Usually a K-shell electron is captured. The mass number remains the same, but the atomic number increases by one because there is an extra proton.

The loss of an electron in the K-shell leaves the atom energetically unstable and so an electron from a higher orbital fills the vacancy in the K-shell. The electron that fills the vacancy is, by definition, losing energy. This excess energy is releases as photons or electrons. Electrons emitted in this way are called Auger electrons (Fig. 2.11). I-125 (used in prostate brachytherapy seeds) decays by electron capture.

2.3.5 Gamma rays

These are packets of electromagnetic rays that originate from the nucleus. Both alpha and beta decay processes may often leave the nucleus energetically unstable, and this excess energy is released as gamma rays, eg Cobalt-60.

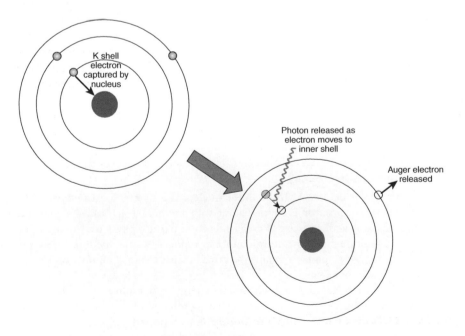

Fig. 2.11 Electron capture producing Auger electrons.

$$^{60}_{27}\text{Co} \rightarrow ^{60}_{28}\text{Ni} + \text{electron} + 1.17 \text{ MeV gamma}$$

$$^{60}_{28}\text{Ni} \rightarrow ^{60}_{28}\text{Ni} + 1.33 \text{ MeV gamma}$$

2.3.6 Internal conversion

In contrast to gamma decay, where an energetic nucleus releases its excess energy as a photon, the energy can be transferred to an orbiting electron, usually a K-shell electron.

The electron then uses some of this energy to escape the atom, and travels with what is left of the energy, as kinetic energy.

The resulting vacancy is filled by an electron from a higher energy level. The cascade of electrons to the lowest overall energy state releases the excess energy as photons or Auger electrons.

2.3.7 Activity and half-life

The activity of a radioactive material is measured as the number of nuclei that disintegrate per second. The SI unit of activity is the becquerel, the symbol is Bq. The unit is named after Antoine Henri Becquerel (1852–1908), a French physicist who won the Nobel Prize in Physics in 1903 as the discoverer of Radioactivity. One of his doctoral students was Marie Curie, after whom another unit of radioactivity, the **curie** (symbol **Ci**) is named. One curie is approximately the activity of 1 gram of radium 226, which decays at the rate of 3.7×10^{10} disintegrations per second i.e. $1 \text{ Ci} = 3.7 \times 10^{10} \text{ Bq}$.

The activity of any radioactive material reduces with time. The activity at any particular time is dependent on the number of nuclei present at that time. The proportion of nuclei undergoing disintegration remains constant. This leads to a pattern of decay called 'exponential decay' (Fig. 2.12).

Half-life is defined as the time for a radioactive material to lose half of its activity, which is the same as saying it is the time for half the nuclei in a material to decay.

The mathematical equation describing the activity at any particular time is:

$$A_t = A_0 e^{-\lambda t}$$

A_t is the activity at the time, t.
A_0 is the activity at time zero.
λ is constant which depends on the half life, T½
$\lambda = 0.6931/ \text{ T½}$

2.3.8 Sources of radioactive materials

Radioactive materials are either naturally occurring or artificially produced.

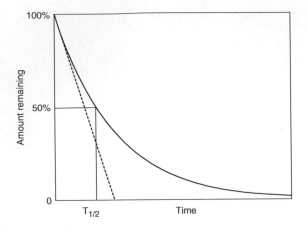

Fig. 2.12 A graph illustrating exponential decay and the half-life.

2.3.9 **Naturally occurring radioactive materials**

There are three naturally occurring radioactive decay series:

- Uranium-238————> (Radium-226)————> Lead-206
- Actinium-235——————————————————> Lead-207
- Thorium-232——————————————————> Lead-208

These naturally occurring radioactive materials tend to have high atomic numbers. The half-life is higher for those with a higher atomic number. They all decay to lead and undergo a series of steps in this decay. For example, uranium, with a half-life of 4.5 billion years emits an alpha particle to become thorium-234, which in turn emits a beta particle to become protactinium-234, and a further 11 steps to become lead-206.

Whilst radium was the first radioactive isotope used for therapy, these are not safe for use today—they have a long half-life, emit gas and decay by alpha emission.

2.3.10 **Artificially produced radioactive materials**

Artificially produced radioactive materials are made in one of three ways

1) Fission,
2) Neutron bombardment,
3) Charged particle bombardment.

2.3.10.1 **Fission**

Fission occurs in a nuclear reactor. It is the splitting of a large atom into roughly two equal parts. It occurs when a neutron enters a nucleus, making the nucleus unstable and results in a neutron that has kinetic energy, the atoms resulting from the fission reaction, and energy.

In nuclear fission, Uranium-235 splits into two separate atoms, one or more neutrons and a lot of energy. There are many different products of nuclear fission, an example of one of the fission reactions is:

$$_{92}^{235}U + {}_0^1n \rightarrow {}_{54}^{134}Xe + {}_{38}^{100}Sr + {}_0^1n + {}_0^1n + Energy$$

The neutrons then react with other uranium nuclei and the reaction continues as a chain reaction.

Fission reactions produce:

- Strontium-90, used for the treatment of bone metastases.
- Caesium-137, formally used in moderate dose rate brachytherapy.
- Tellurium-131 decays to iodine-131, used in imaging and treatment of thyroid disease.

2.3.10.2 Neutron bombardment

Here, a stable element is placed in a nuclear reactor, bombarded with neutrons and the nucleus of the stable element captures a neutron. This leads to rearrangement of the nuclear components, and the release of energy in the form of gamma rays.

Cobalt-60 used in many external beam therapy machines around the world is produced in this way. Other useful products of neutron bombardment are phosphorus-32, used in the treatment of polycythaemia, and molybdenum-99 that decays to the metastable isomer, technetium-99m which is used for bone scans.

2.3.10.3 Charged particle bombardment

This is when a stable element is bombarded with protons or alpha particles, leading to absorption of the particle and ejection of one or more neutrons. The resulting element has a higher atomic number, i.e. a proton gain. Iodine-123 is produced in this way, used in the imaging, and some treatments, of thyroid disease.

2.3.11 Radioactive equilibrium

Radioactive equilibrium occurs when the rate of production of a radioisotope (from its parent) is equal to its rate of decay, and so its quantity remains constant. For radioactive equilibrium to occur, the half-life of the parent has to be greater than the half-life of the daughter product.

An example is in the production of technetium-99m, the daughter product of molybdenum-99.

2.3.12 Relevant key points in radioactivity

Beta minus decay products are the most commonly used in medicine. They are produced in a nuclear reactor by either fission production or neutron capture. They are relatively cheap and have a long half-life.

Examples in clinical practice are cobalt-60, iridium-192, phosphorus-32, technetium-99, caesium-137, strontium-90, and iodine-131.

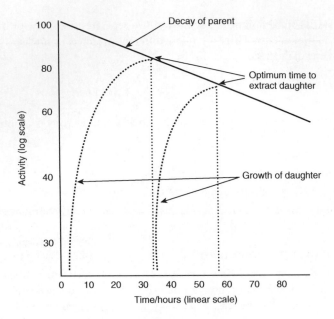

Fig. 2.13 Diagram of radioactive equilibrium.

Beta positive decay products are produced by charged particle bombardment in a cyclotron. They are costly and have a short half-life. Examples are carbon-11 and fluorine-18, both of which are used in imaging.

Chapter 3

The life of a photon

A Sibtain, A Morgan and N MacDougall

3.1 Introduction

X-rays were discovered by Wilhem Conrad Roentgen in the late afternoon of 8th November 1895. Within two weeks he had taken an X-ray image of his wife's hand, and by the end of the year he published the paper, 'On A New Kind Of Rays' (Über eine neue Art von Strahlen). The discovery earned him the Nobel Prize in 1901, around the time of the discovery of radium. He called them X-rays because he didn't know what they were; in algebra, X denotes an unknown, so he called them X-rays . . . and it stuck.

3.2 The birth of a photon: X-ray production

This section provides the basis of understanding the production of photons in a linear accelerator.

X-rays are electromagnetic radiation at the high energy end of the electromagnetic spectrum (see fig 2.9). They are formed when electrons travelling at high velocity interact with other electrons or nuclei in matter that has a high atomic number. These interactions can be collisions or close encounters. The electron loses energy from these interactions, and this energy is released as electromagnetic waves: the photon is born!

There are a number of possible interactions the travelling electron can have:

(1) Interactions with orbiting electrons, either
   (a) Outer orbiting electrons
   (b) Inner orbiting electrons
(2) Interactions with the nucleus
(3) Collisions with the Nucleus

3.2.1 Interactions with the orbiting electrons

3.2.1.1 Outer orbiting electrons

The travelling electron can impart energy to an orbiting electron which can gain enough energy to escape from the nucleus it is orbiting. This interaction is called **ionization** but in this particular case, a photon is not produced.

The amount of energy delivered may be not enough to cause ionization, but only enough to cause the electron to vibrate. The travelling electron continues on its journey, in a different direction, with less kinetic energy.

3.2.1.2 Inner orbiting electron

The travelling electron interacts with an inner shell electron (either a K or L shell—[see Chapter 2]). The energy transferred allows the inner shell electron to escape the atom, i.e. ionization. The vacancy in the K or L shell is filled by an electron from a M or N shell. The electrons in the more distant shells have a higher energy. When dropping to a lower energy level they release their excess energy as electromagnetic radiation, i.e. photons. The energy released is the difference between the binding energy of the two shells the electron travelled between.

3.2.1.2.1 Characteristic X-rays

The electron shells in a particular atom have their own particular binding energy. The binding energy of each shell depends on the atomic number of the element. The larger the nucleus is, the greater the binding energy of the electron shells. The set of binding energies are particular to each element and varies between different elements. Therefore when an inner shell electron is removed from any particular atom and a higher shell electron fills that vacancy, the energy of the resulting photon is also particular to that element. The range of photon energies that emanate from that element are the characteristic X-rays of that element.

3.2.1.3 Interactions with the nucleus

The nucleus carries a positive charge and the electron carries a negative charge. The nucleus is also substantially larger than the electron. A passing electron travelling at high speed near a nucleus will be influenced by this attraction, and as a result, will change direction and lose some of its kinetic energy. This 'lost energy' is released as electromagnetic energy, i.e. a photon. The electron will have slowed down, as though it has 'put the brakes on'. The photons/X-rays/electromagnetic radiation (all the same thing) released in this way are called 'braking radiation', more usually using the German—Bremsstrahlung.

This mechanism of X-ray production produces a wide range of energies across the X-ray spectrum, ranging from very low up to the maximum kinetic energy of the incoming electrons.

The intensity of bremsstrahlung directly depends on:

1. The size of the interacting nucleus, i.e. the atomic (Z) number,

2. The kinetic energy of the incoming electrons.

Bremsstrahlung is the dominant way in which clinical X-rays are produced when high atomic number materials irradiated with electrons.

3.2.1.4 Collisions with the nucleus

This is a rare event for any charged particle, particularly electrons, and as such is outside the scope of this book.

3.2.2 Which way do the photons go?

The direction the newly formed photons take after production is discussed in more detail in Chapter 11. Referring to Fig. 11.2, it can be seen that at low kV energies

(~100kV), photons produced travel pretty much in all directions or, in other words, isotropically. As the energy of the electron beam increases, the direction of photon travel starts to be biased towards the direction of the original electron beam—or what is termed "the forward direction." Once in the MV energy range, the majority of photons are produced in the forward direction. This phenomenon influences the design of targets used in therapy equipment. At low kV energies, the photon beam emerges at an angle of 90° to the electron beam but at MV energies, the photon beam emerges from the target in the same direction as the electron beam.

3.2.3 The X-ray Spectrum

When X-rays are produced by electrons entering a target, a range of photon energies is produced. If drawn as a histogram with the frequency of each energy on the y axis (the same as the intensity) and the energy value on the x axis, the result is shown in Fig. 3.1.

There are two parts to the X-ray spectrum:

1) There is a *continuous spectrum* that results from bremsstrahlung. The continuous spectrum depends on the energy of the incoming electrons and on the atomic number of the target material.

2) There are spikes of particular energy intensities that are particular to the target material, and are the *characteristic* X-ray part of the spectrum.

3.3 The basic components of an X-ray tube: how to bring the photons to life

To bring photons to life we need an instrument that can:

♦ Generate electrons,

♦ Move these electrons at high speed (i.e. give them kinetic energy),

♦ Direct them at a high atomic number target.

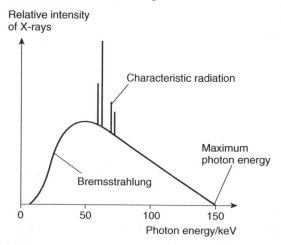

Fig. 3.1 An X-ray spectrum.

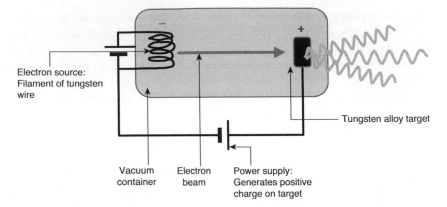

Electron source:
Filament of tungsten
wire

Tungsten alloy target

Vacuum   Electron   Power supply:
container   beam   Generates positive
                          charge on target

Fig. 3.2 The basic components required for X-ray production.

The basic components of this instrument—the X-ray tube (Fig. 3.2)— are therefore:

- An electron source to generate the electrons,
- A positively charged target in which the electrodes can interact,
- A vacuum container to give the electrons a clear path to the target,
- A high voltage power supply.

3.3.1 **The electron source**

This is a thin filament of tungsten wire that has a high electrical current passed through it, i.e. lots of electrons flowing through the wire. The electrical current is high, so the filament heats up and some electrons escape the wire and form a cloud of electrons around it. This process is known as thermionic emission.

The filament current can be varied which in turn varies the number of electrons emitted into the electron cloud.

3.3.2 **The target**

The target needs to contain a high atomic number material to give plenty of opportunity for bremsstrahlung to occur. It also needs a high melting point and to be designed to conduct heat away effectively. It is usually made of a tungsten alloy.

3.3.3 **The power supply**

The power supply generates the positive charge on the target. The higher the voltage in the system, the greater the difference in charge between the negatively charged electron source (the cathode) and the positively charged target (anode).

The voltage can be varied. This will vary the energy of the electrons being accelerated into the target, which in turn varies the energy of the X-rays produced.

These basic components are similar for all X-ray machines, from diagnostic to therapy. The details of a linear accelerator are given in Chapter 11, Beam Therapy Equipment.

3.3.4 **Adjusting the intensity and quality of the X-rays**

The intensity of X-rays is an expression of the number of photons passing through a unit area. It is a measure of the amount of energy flowing.

An increase the number of photons being produced requires an increase in the number of electrons produced and in the number of electrons attracted from the electron cloud to the target. These are achieved by increasing the filament current and the tube current respectively.

3.3.5 **X-ray beam quality**

The quality of an X-ray beam describes its penetrating power. It depends on the energy of the photons in the beam and therefore depends on the energy of the electrons entering the target. The quality of the beam is therefore increased by increasing the tube voltage.

3.4 **Photon interaction with matter: the beginning of the end**

The photon interactions described above discuss the interaction at an atomic/single photon level. A photon beam however is a continuous stream of photons. The interaction of the beam with matter is the focus of this section.

When a photon beam leaves a linear accelerator it has a particular intensity, called fluence. The fluence is the number of photons passing through a sphere in space (Fig. 3.3). The greater the fluence, the greater the number of photons. The energy fluence is the total energy carried by these photons as a whole.

The intensity of a beam reduces as it passes though a material. This reduction in intensity is called attenuation.

There are two processes at play in photon beam attenuation:

A) Absorption—when a photon gives up all of its energy to the material by transferring it to an electron.

B) Scatter—the photon collides with an electron or atom and changes direction with or without a change in energy.

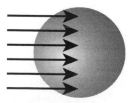

Fig. 3.3 The intensity of a photon beam is a measure of the number and total energy of photons passing through a sphere in space.

There are thought to be up to twelve ways in which photons interact with matter of which only three are of relevance to the radiation oncologist.

When a photon enters matter, there are three possible things that can occur. Which of these occurs depends on the energy of the incident photon and the matter with which the photon is interacting.

The interactions are called:

♦ Photoelectric interaction,

♦ Compton scatter,

♦ Pair production,

These interactions all compete with each other, but the probability of any one of them happening depends on, to varying degrees:

♦ The energy of the photon,

♦ The atomic number (Z) of the material,

3.4.1 Photoelectric interaction

Photoelectric interaction is similar to the 'inner electron interaction' that occurs when an electron interacts with an atom's inner shell electron, described above. The difference here is that it is a photon, rather than an electron, that gives up all its energy to an inner shell bound electron. The energy given to the electron can allow it to escape the binding energy of the nucleus, and the vacancy is filled with an electron from another, higher energy shell. The energy released by the electron that fills the vacancy is released as a photon, which carries energy equal to the difference in the binding energies between the two shells it travels between (Fig. 3.4). *This energy is the characteristic radiation of the material.*

The probability of photoelectric interaction is:

♦ Directly proportional to the atomic number cubed (Z^3),

♦ Inversely proportional to the energy of the photon cubed (E^{-3}).

Therefore, photoelectric interaction increases very rapidly with the atomic number of the material and decreases very rapidly with the energy of the photons.

So with low energy photons the photoelectric effect is predominant, and the interaction leads to most of the energy of the incoming photons being absorbed.

The characteristic radiation energies vary from material to material:

♦ Carbon   0.3 keV

♦ Oxygen   0.5 keV

♦ Calcium   4 keV

♦ Lead   80 to 88 keV

Practically, the photoelectric effect is most relevant in diagnostic radiology, leading to contrast between tissues of different densities and atomic numbers—those with higher density and atomic numbers, e.g. bone, containing calcium, are more likely to interact with the incoming low energy photon, absorbing its energy, compared to soft tissue which is predominantly made up of water.

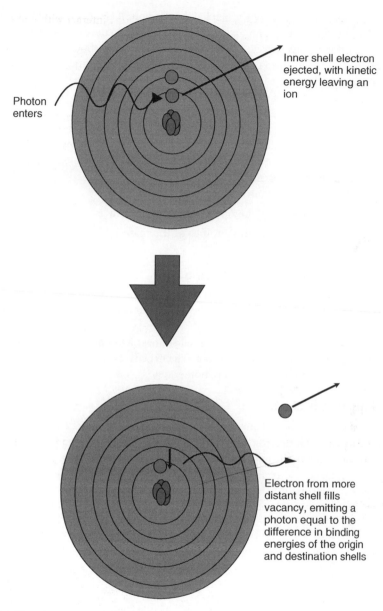

Fig. 3.4 The photoelectric effect. This figure is reproduced in colour in the colour plate section.

3.4.2 **Compton Interaction**

Compton interaction, also called Compton scattering, is named after the American Physicist, Arthur H. Compton. **It is the most relevant interaction that occurs in the energies used in radiation therapy.**

It can be thought of as a snooker ball like interaction. A travelling photon collides with an electron. Some of the photon's energy is transferred to the electron. The electron then moves off with this energy. The photon continues on with less energy, but in a different direction. The greater the angle by which the photon's direction changes, the greater the energy transferred between the photon and the electron.

The angle at which the photon direction changes can range from a very slight 'glancing blow' at around 0°, to a head on collision at 180°. The electron moves off at any angle, Φ, between 0 and 90°. Increasing the photon energy results in more electrons and photons being scattered in the forward direction, as the 'glancing blow' is more likely.

Figures 3.5 and 3.6 show how the angular distribution of the scattered photon and the scattered electron varies with different initial photon energies.

The probability of Compton interaction:

- Decreases with increasing photon energy in the MV range
- Is independent of the atomic number of the material in which the photon is interacting.

3.4.2.1 Elastic and inelastic scattering

Compton scatter/interaction is also called *inelastic scattering*, because the total energy is conserved. The original photon energy is redistributed between itself and the electron it interacts with.

Elastic scattering occurs when a photon passes near a bound electron and causes it to vibrate and give out low energy photons. The net effect is that the material takes up no energy. It is an irrelevant effect in radiotherapy.

3.4.3 Pair production

This occurs when a photon changes from a packet of electromagnetic wave energy into a pair of particles—an electron and a positron. All the energy carried by the incoming photon is converted to the mass and kinetic energy of these particles.

The pair of particles then deposits their energy in the material. The positron interacts with an electron and the mass of the two combining particles turns into two photons. The energy of these photons is 0.511MeV each. Pair production can only

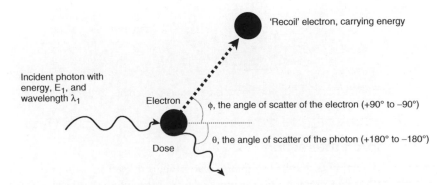

Fig. 3.5 Compton scatter.

$$\lambda_2 - \lambda_1 = \Delta\lambda = h/mc(1-\cos\theta)$$

λ_1 The wavelength of the incident photon

λ_2 The wavelength of the scattered photon

h   Planck's constant

m   The mass of an electron

c   The speed of light

$\cos\theta$ The cosine of the angle of scatter of the photon

The difference in the wavelength ($\Delta\lambda$), and therefore the energy, between the incident and the scattered photon is equal to the energy given to the electron. This energy is inversely proportional to one minus the cosine of the angle of the scattered photon. This works out so that the energy is the most it can be if the photon bounces straight back.

Fig. 3.6 An explanation of the Compton equation.

occur if the energy of the incoming photon is greater than 1.022MeV (the combined mass of the electron and positron created).

The probability of pair production is:

- Directly proportional to the atomic number of the material,
- Directly proportional to the energy of the photon.

3.4.4 **The relative importance of the three main interactions**

When photons interact with matter, there are three possible interactions as discussed above. Which of these occurs depends on two things.

- The energy of the incident photon,
- The atomic number of the material being interacted with.

The probability of each interaction is denoted by a Greek letter:

τ for the photoelectric effect,

σ for the Compton effect,

κ for pair production.

Fig. 3.7 shows the relative importance related to the atomic number of the absorber and the energy of the incoming photons.

3.5 **Attenuation: the exponential attenuation and the attenuation coefficient**

As a beam of photons enters a material it is attenuated.

The intensity emerging from the other side of the absorber is less than that entering the absorber.

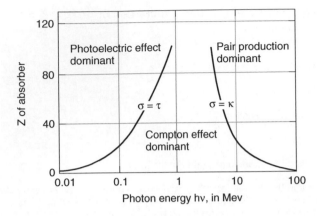

Fig. 3.7 The relative importance of the photoelectric effect, the Compton effect and pair production related to the Z of the absorber and the incident photon energy.

The loss of intensity is proportional to the degree of intensity entering the material and it is proportional to the thickness of the absorber.
This can be written as:

$$\Delta I \propto -x.I_0$$

The elements of this formula are as follows:

ΔI is the change in intensity after passing through material,

I_0 is the intensity of the photon beam at the point of entering the material,

x is the thickness of material.

As with all equations where two components are proportional to each other, a constant can be applied:

$$\Delta I = -\mu.x.I_0$$

where μ is the total linear attenuation coefficient of the absorber and has the unit m^{-1}. It is the fractional reduction in intensity of a photon beam per unit length of the material. The equation assumes x is very small so for greater thickness the expression is integrated:

$$I_x = I_0.\exp(-\mu x)$$

where I_x is the intensity transmitted after the beam has passed through the material of x thickness.

We can derive the mass attenuation coefficient by dividing μ by the density of the absorber, ρ. The mass attenuation coefficient, μ/ρ, has units of $m^2 kg^{-1}$. This gives the attenuation per unit mass rather than per unit path length.

3.5.1 **Polyenergetic beams and broad beam geometry**

The equation holds true for a thin beam consisting of photons of a single energy. In 'real life', photon beams have a range of energies: the value of μ changes with the energy because the lower energy photons are preferentially absorbed earlier in the beam's journey through the material. The effect is to reduce the intensity more than expected i.e. there is more attenuation.

Furthermore, the beams are broad, meaning there is greater scatter of photons as the beam passes deeper into the material. This has the effect of reducing the intensity *less* than expected i.e. there is less attenuation and more of the beam energy gets through than expected.

3.5.2 The half value layer

The half value layer (HVL) is the thickness of a material required to reduce the intensity of a beam by half.

Taking the equation defined above:

$$I_x = I_0 \exp(-\mu x)$$

If the initial intensity, I_0, is to be reduced by half =>

$$I_0/2 = I_0 \exp(-\mu x)$$

X is the thickness of the material needed to reduce the intensity by half, i.e. is the half value layer (HVL):

$$I_0/2 = I_0 \exp(-\mu.HVL)$$

Rearranging the equation gives:

$$HVL = \ln2/\mu$$

Which calculates to:

$$HVL = 0.693/\mu$$

The HVL is used to compare how penetrating the photon beams are. The more penetrating the beam, the greater the HVL. A penetrating beam is said to be 'hard'; it also serves as a description of the quality of the beam. Higher quality beams are harder and have larger HVLs, which really relate to the overall energy of the beam.

The concept of the HVL can be applied to any factor, e.g. the tenth value layer being the thickness of material required to reduce the intensity of a beam to one-tenth of its original value.

3.5.3 Mass energy transfer and mass energy absorption coefficients

Above, we explained the linear attenuation coefficient (μ) and mass attenuation coefficient (μ/ρ). They are particular to whatever material is attenuating a photon beam and have been defined for a wide range of materials e.g. lead, aluminum, bone etc. These coefficients also vary with energy, given that at different energy levels, different processes of photon—electron interaction occur (Fig. 3.7). They relate to the reduction in *intensity* of a beam of photons. The mass energy transfer coefficient and mass energy absorption coefficient describe something similar, but relate to a reduction in the *energy* of a beam of photons, describing the energy lost by the beam and the energy gained by the absorbing material respectively. This concept is revisited in Chapter 7, Radiation Dosimetry.

3.6 **Photon offspring: the secondary electrons and their interactions**

The photon interactions discussed above produce electrons (pair production) or impart energy to them. The electrons will, in turn lose this energy in processes exactly equivalent to the electron interactions that occur in X-ray production:

- ◆ Ionization of an outer orbital electron,
- ◆ Ionization of an inner orbital electron,
- ◆ Production of Bremsstrahlung (negligible in the human body).

These interactions can be thought of as either collisions—ionization of orbital electrons—or radiative interactions—Bremsstrahlung.

An electron, either given energy from a photon or created through pair production, will follow a path though the material. It will lose energy along the way though the processes, until it stops. The point at which the electron stops varies according to the material through which it is travelling, and the energy the electron carries. Four features of this path are defined below.

3.6.1 **The path length**

This is the absolute distance travelled by the electron as it passes thought the material.

3.6.2 **The range**

This is the distance in the material over which energy is deposited. The range is effectively the depth of maximum penetration of a charged particle beam. For example, the path length and the range are very similar for all protons in a monoenergetic beam as they don't undergo wide angle scattering and exhibit a pronounced Bragg peak (see Chapter 4), whereas electrons in a monoenergetic beam undergo lots of wide angle scattering; while the electrons' path lengths might be similar, the range of the beam is the depth of maximum penetration The range is dependent on the material and the energy the electron carries.

3.6.3 **The stopping power**

The 'stopping power' describes the material's ability to stop an electron (or indeed any charged particle) and is the rate at which the electron loses energy along its track. It relates to both the collision type and the radiative type interactions.

The unit of stopping power is Joules per metre ($J\,m^{-1}$).

The stopping power divided by the density of the material, gives the stopping power per unit mass of the material ($J\,m^2\,kg^{-1}$).

3.6.4 **The linear energy transfer (LET)**

The linear energy transfer is the rate at which energy is deposited along a particle track. It is particle dependent, and is different to stopping power. LET defines the amount of energy deposited in the material surrounding the particle track as opposed to the amount of energy lost by the particle. Linear energy transfer therefore relates to the collision type interactions only. LET is usually expressed in units of keV per micrometre (keV μm^{-1}).

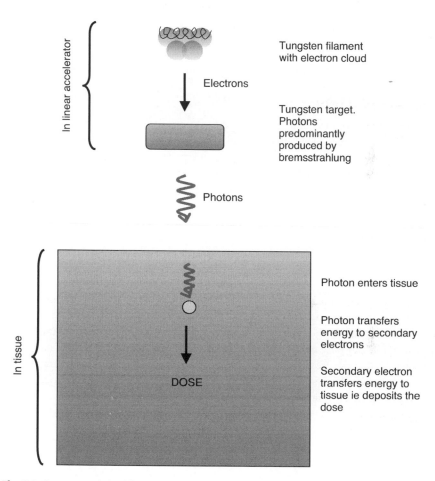

Fig. 3.8 Summary of the life cycle of a photon.

Chapter 4

Electrons, protons and neutrons

A Morgan

4.1 Introduction

The vast majority of radiotherapy treatments are performed using megavoltage photons, the physics of which have been discussed elsewhere in this book. However, the photons interact with electrons, transferring energy to them and it is these electrons which then move through tissue causing the majority of biological damage.

Electron beams have been used in radiotherapy for many years. They can be used to treat superficial tumours without damaging potentially normal tissues lying at deeper depths (see chapter 9).

There is now considerable interest in the use of proton beams and carbon ion beams in the treatment of cancer. Their unique depth dose characteristics result in the irradiation of less normal tissue than the most conformal photon treatments.

Neutron beams have been used in the past but fell out of favour due to unacceptable (and unexpected) side effects. However, they can be present in high energy photon beams and the interactions they undergo are worthy of consideration.

Electrons, protons and carbon ions are all charged particles. Neutrons are uncharged particles but can interact with nuclei in a material to produce potentially damaging charged particles so are of indirect interest in this chapter. A proton is just a hydrogen nucleus with its electron removed. All six electrons must be removed from a carbon atom to completely ionize it.

Charged particles have an electric field associated with them and it is because of this field that charged particles lose energy when they travel through a medium. This lost energy is absorbed in the medium and can damage cellular DNA, leading to cell death—and so we have a potential means of treating tumours. . . . but we're getting ahead of ourselves. We need to look at the interaction processes in a little more detail.

Charged particles lose energy in a different manner from that of uncharged radiations: X-rays, gamma rays and neutrons. Photons and neutrons may pass through matter with no interactions and therefore no energy loss. When they do lose energy, they tend to lose it in one or more discrete events such as a series of Compton type interactions or a photoelectric absorption.

Charged particles, however, possess an electric field and they interact with the electrons or with the nucleus of virtually every atom which they pass. The probability of an uncharged particle passing through a layer of matter without interaction may be small but is finite. The probability of a charged particle passing through a layer of matter without interaction is zero. A single 10 MeV electron may undergo around a million separate interactions before losing all of its kinetic energy.

The most important type of interactions which occur between a charged particle and an atom are illustrated below using a simple model. In this model, electrons exist in stable states of energy characterized by discreet levels surrounding the nucleus of an atom as discussed in Chapter 2.

4.2 Types of charged particle interactions

Charged particles can lose energy by several processes when they pass through a medium. The mode of interaction is determined largely by the energy of the incident charged particle and by its distance of approach to the atom or nucleus. Consider the radius of the atom to be a. The distance of closest approach to the nucleus is b (see fig 4.1).

4.2.1 Soft collisions (where b is greater than a)

If a charged particle passes an atom at a distance which is large compared to the size of the atom, there are two possible interaction mechanisms:

i) Excitation of atomic electron to a higher level which returns to the ground state with emission of a photon.

ii) Ionization of atom by excitation of valence shell electron. The net effect is to transfer of a few eV of energy to the medium.

It can appreciated that large values of b are more probable than 'hits' on individual atoms and therefore soft collisions are the most numerous type of charged particle interaction.

The absolute velocity of light in a vacuum is 3×10^8 metres per second. However, the velocity of light in a material such as water is less. While nothing can travel faster than the speed of light in a vacuum, it is possible for a highly energetic charged particle to travel at a velocity greater than light in certain media. If such a medium is transparent

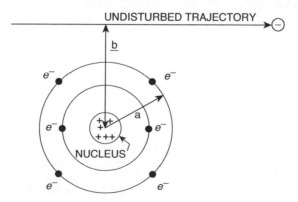

Fig. 4.1 Relationship of impact parameter to atom.
Adapted from Attrix, F. H. (2004) 'Introduction to Radiological Physics and Radiation Dosimetry' Fig. 8.1, p. 161. London: Wiley, VCH.

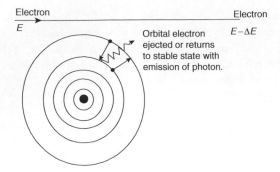

Fig. 4.2 Simple representation of soft collision.
Adapted from Klevenhagen, S C (1985) *'Physics of Electron Beam Therapy'*, Fig 2.1a,
p. 38. Bristol: Adam Hilger.

then a very small part of the energy spent in soft collisions is emitted in the absorbing
medium as a coherent bluish white called **Cerenkov radiation.** The quantity of energy
lost by Cerenkov radiation is less than 0.1% that lost by 'soft' collisions.

Cerenkov radiation has been cited as a possible reason for visual disturbances occasionally reported by patients undergoing radiotherapy close to the eye. High energy
electrons may pass through the vitreous humour, resulting in Cerenkov production.

4.2.2 Hard Collisions (b is roughly equal to a)

When b is approximately equal to a, it is more probable that the incident charged
particle will interact with one of the orbiting atomic electrons which is then ejected

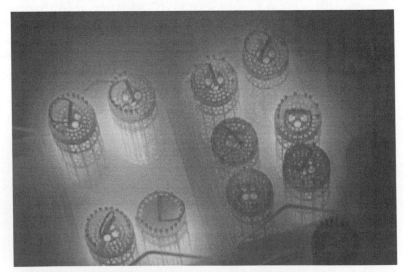

Fig. 4.3 Nuclear reactor rods in a water filled cooling pond producing Cerenkov
radiation. This figure is reproduced in colour in the colour plate section.
Image courtesy of Nordion Inc.

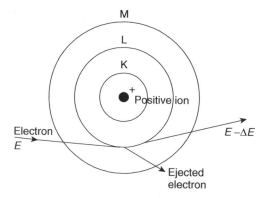

Fig. 4.4 Representation of hard collision.
Adapted from Klevenhagen, S C (1985) *'Physics of Electron Beam Therapy'*, Fig 2.1b,
p. 38. Bristol: Adam Hilger.

from the atom with significant kinetic energy. This process is known as a hard collision.
The ejected atomic electron is known as a delta-ray (δ-ray). Even though hard collisions
are few compared to soft collisions, the fraction of energy lost by hard collisions is
large and overall, is roughly the same as that lost by soft collisions (see fig 4.4).

If an inner shell atomic electron is ejected, characteristic X-rays or Auger electrons
are emitted which can result in energy being deposited away from the primary particle
path.

An Auger electron can be considered to be the result of a characteristic X-ray being
captured by an orbital electron which is then ejected from the atom.

4.2.3 **Radiative interactions with the nuclear electric field**

If b is very much less than a, coulomb interaction takes place mainly with nucleus and this
is most important for incident electrons. In the vast majority of cases (more than 95%),

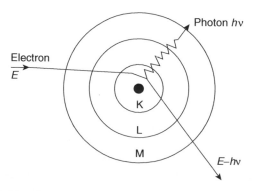

Fig. 4.5 Radiative interaction with nuclear field.
Adapted from Klevenhagen, S C (1985) *'Physics of Electron Beam Therapy'*, Fig 2.1c,
p. 38. Bristol: Adam Hilger.

the interaction with the nucleus is considered to be elastic. The electron's path will be changed and it may lose a small but negligible amount of energy. Elastic nuclear interactions give rise to changes in the direction of the impinging electron but not to significant energy losses, so this is not a mechanism for transfer of energy to a medium. It is however an important means of scattering electrons and is the main reason why electrons follow tortuous paths, particularly in materials with a high atomic number (see Fig 4.5).

In the other 5% or less of cases, an inelastic interaction occurs. A charged particle passing near a nucleus may be deflected from its path by the action of the nuclear Coulomb force and lose energy. As the charged particle passes the vicinity of the nucleus, it suffers deflection and deceleration. As a result, part of its kinetic energy is dissociated from it and appears as a photon. The charged particle may give up to 100% of its kinetic energy to the photon. This process is known as bremsstrahlung—German for 'breaking radiation'. The emitted radiation covers the entire spectrum of energies up to the maximum kinetic energy of the charged particle. The likelihood of a radiative interaction taking place depends on the square of the atomic number of the irradiated material (Z^2). It also depends inversely on the mass of the irradiating particle and so is considered insignificant for all charged particles, except electrons.

It should be noted that most photon beams used in radiotherapy are generated by accelerating electrons to a high velocity and then stopping them suddenly (see Chapter 3 'The life cycle of a photon').

4.3 **Stopping Power**

The rate at which a charged particle loses energy as it passes through a medium is known as the **stopping power**. The units of stopping power are given as Joules per metre (J m^{-1}) as is usually shown by the term dE/dx—the rate of energy loss with distance travelled.

4.3.1 **Variation of dE/dX with X**

A typical graph of dE/dX against distance travelled, X, for a charged particle is shown in Fig. 4.6

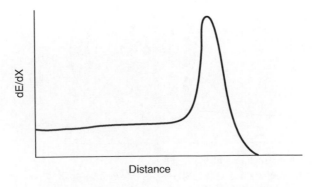

Fig. 4.6 Variation of energy loss with distance for a charged particle.

This type of graph is representative of the pattern of energy loss of every charged particle. It is characterized by a relatively low and constant rate of energy loss immediately after entering a medium. However, towards the end of its path, the rate of energy loss rises dramatically and falls to zero. This peak in the curve is known as the Bragg peak, after its discovery by William Henry Bragg in 1903.

The mathematical equation which describes the shape of the curve in Fig. 4.6 is rather complex but there are a few important points which can be clearly stated. The rate of energy loss with distance (dE/dX) is:

- *Proportional to the square of the charge on the particle* . . . so an alpha particle, which has a charge of +2, will lose energy four times as fast as a proton.

- *Inversely proportional to the square of velocity of the charged particle* . . . as the particle slows down, the rate of energy loss increases—which agrees with the shape of the graph above.

- *Independent of the mass of the charged particle*—this means that for particles of the same velocity, the rate of energy loss of a proton is similar to that of an electron as both have a charge of 1 unit.

Fig. 4.7 shows the pattern of absorbed dose from an electron beam and a proton beam. The characteristics appear to be completely different but the energy loss of each particle type is exactly the same as shown above. The reason relates to the relative mass of each particle. Electrons will undergo interactions with other electrons—which are

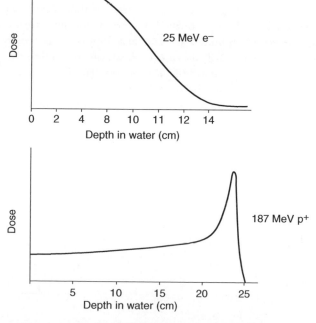

Fig. 4.7 Variation of dose with depth for electron and proton beams.

the same mass—and will be easily scattered by the interactions undergone. Many will end up travelling the direction they have come from—hence there is no well defined Bragg peak for the beam as a whole. Protons on the other hand, having a mass nearly 2,000 times greater than an electron are less easily deflected and an obvious Bragg peak is seen. Consider an electron as a ping pong ball and imagine firing a ping pong ball into a collection of other ping pong balls—the original is unlikely to travel through the collection without deflection from its path. If you consider a proton as a ten pin bowling ball and fire that at the same collection of ping pong balls, it is easy to imagine that it would plough through them with minimal path deflection.

4.3.2 Restricted stopping power

It has been shown that charged particles generally lose energy in a large number of small interactions with small energy losses via soft collisions. This has given rise to the description of electron motion in terms of a 'continuous slowing down approximation' (CSDA). The energy transferred to the medium in this way may be assumed to be absorbed locally, i.e. within a small volume close to the point of interaction. In this case, it is usually safe to assume that the energy lost by the charged particle is the same as that absorbed locally.

Remember δ-rays? These are energetic electrons (and so not really rays) resulting from large energy transfers to an atomic electron via hard collisions. The atomic electron is able to travel a distance and produce ionization far from its point of origin. Delta-rays may have ranges comparable with those of the primary charged particles which create them.

The concept of restricted stopping power is necessary to draw attention to the energy lost by electrons that is absorbed in close vicinity to the electron path rather than on the total energy dissipated by the electron. The dose deposited by a charged particle in a given area may be overestimated unless δ-ray equilibrium exists . . . i.e.—for every δ-ray that leaves a small volume of material, a δ-ray enters the volume to replace the energy lost. This is not usually the case. In short, energy lost by a charged particle cannot necessarily be considered to be equal to energy deposited locally . . . and energy deposited per unit mass is what we know as dose.

In radiobiology, restricted stopping power is known as **linear energy transfer** (LET) and represents the stopping power for all collisional interactions, including the production of δ-rays, up to a specified cut off value. Interactions from radiative interactions are

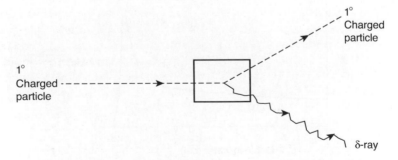

Fig. 4.8 Energy loss from a small volume due to production of a delta-ray.

ignored as it is assumed that the photons produced will interact with another electron a long way from the radiative interaction site.

Ionizing radiation interacts with matter in a similar way but different types of radiation differ in their effectiveness in damaging a biological system. The most important factor that influences the relative biological effectiveness of a type of radiation is the distribution of the ionizations and excitations in its path.

LET is used to describe both excitation and ionization events. Given the shape of the curve shown in Fig. 4.6 above, it can be appreciated that the LET of a particle will change with distance travelled. Commonly, LET values quoted are an average of energy lost with distance. The LET gives the average energy loss of a particle per unit length of travel in terms of keV/micrometre, keV/µ. The variation of energy loss along the track of a charged particle has led to the utility of LET being questioned. However, the fact remains that it is a valuable method of comparing the energy deposition characteristics of different radiation modalities.

LET values vary from publication to publication, but the following may be taken as representative. All values quoted are in terms of keV/µ. An electron beam with a kinetic energy of 10keV will have an LET of 2.5, whereas an electron beam with an energy of 1 MeV will have an LET of 0.2. Remember, the greater the energy, the greater the distance travelled, so the average for a given particle type will decrease with increasing kinetic energy. Proton beams of energy 10 MeV have an LET of around 5 while those of 100MeV have a value of around 0.5.

Alpha particles are relatively large and slow moving, so they will lose energy more quickly both because of the velocity and charge dependence on dE/dX. A 5.3 MeV alpha particle, such as emitted by Polonium-210 has an LET of almost 50.

Neutrons are not directly ionizing themselves but they may cause a nucleus to break up leading to the production of heavy charged nuclear fragments, with a correspondingly high LET.

The high LET exhibited by alpha particles is the main reason why the absorption of alpha emitting isotopes into the human body is of great concern. They cause a great deal of damage to normal tissues and have been linked with the development of bone tumours.

Radium behaves in a similar way to calcium when ingested, and is readily absorbed by bone where it may sit, irradiating bone marrow and other tissues. In 1917 the U.S. Radium Corporation started producing a radium containing paint called Undark, which as the name may suggest glowed in the dark. The company employed several thousand employees, mainly women, to paint Undark onto the hands and dials of watches. The employees were encouraged to keep the lines and characters they painted sharply defined by licking the tips of their brushes, thus continually ingesting small amounts of radium on a regular basis. Large numbers of the workers developed serious health issues including anaemia and jaw bone necrosis resulting in tooth loss. Significant numbers went on to develop tumours.

4.4 Neutrons

Neutrons are not charged and hence are not directly ionizing but they interact quite readily with nuclei and can set protons and other nuclear fragments in motion by knock-on collisions. Photon interactions with matter almost always result in the production of high energy electrons but neutron interactions with matter are not readily categorized. There are several outcomes of neutron interactions but generally two processes are likely.

Elastic scattering. A neutron interacts with a nucleus as whole. The nucleus gains kinetic energy and recoils through the medium. The original neutron loses energy and is deflected from its original path. The transfer of energy is greatest when the target nucleus is lightest i.e. for hydrogen atoms.

Inelastic scattering. Inelastic scattering is considered to occur when a neutron is absorbed by a nucleus, rather than scattering off it. This is where things start getting a little complex. The nucleus will be unstable and several different phenomena may occur to return it to a more stable state. It may eject one or more neutrons, which can then go on and interact with other nuclei. It may eject a proton, alpha particle or larger nuclear fragment with high LET, depositing considerable energy and causing considerable normal tissue damage in the human body.

The nucleus may also eject a high energy photon in order to return to a lower energy state.

Following some unsuccessful clinical trials in the 1970s and 1980, neutron beams are no longer used. However, clinical photon beams may be contaminated with neutrons. How so?

Clinical photon beams are produced by stopping high energy electron beams. The higher the energy of the electron beam, the higher the energy of the resulting photon beam. As discussed in an earlier chapter, protons and neutrons are bound together in a nucleus and that binding energy is around 8MeV per nucleon. If a photon with energy higher than 8 MeV is absorbed by a nucleus, a neutron can be ejected.

This can be a problem in the radiotherapy department. As highlighted above, it is important for staff safety that treatment rooms containing high energy photon treatment units are adequately shielded against neutron leakage. The emission of a high energy photon following neutron absorption does not happen instantaneously. This 'induced radioactivity' may persist for several minutes after a high energy photon treatment has finished, meaning that staff entering the treatment room may be exposed to a low intensity high energy photon field, posing a radiation protection issue.

The problem is more evident if a technical problem requires disassembly of the treatment head. The induced radioactivity in the vicinity of the linear accelerator target may mean that the commencement of service work may be delayed for several hours so as to allow dose rates to service personnel to reduce to acceptable levels.

4.5 **Principles of heavier charged particle therapy**

As discussed earlier, the rate of energy loss of all charged particles exhibit a characteristic shape, terminating in a Bragg peak.

Protons transfer energy to a medium in the same ways as electrons. However, being around 2000 times heavier, they are not easily deflected from their paths and so retain a distinctive Bragg peak at depth. The highly localized deposition of dose with minimal irradiation of normal tissues is a highly prized goal in radiotherapy and proton beams

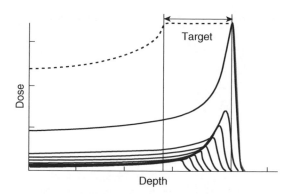

Fig. 4.9 Spreading out a proton Bragg peak to cover a target.
Adapted from www.aapm.org/meetings/03SS/Presentations/Lomax.pdf with permission
from Dr Tony Lomax.

are of considerable use in this respect. However, for a monoenergetic proton beam, the Bragg peak can be so narrow that it is not clinically usable. While we aspire to miss normal tissues, missing the target is not an option.

There are two main ways by which this may be achieved and this depends on which method is used to produce the proton beams.

From Fig. 4.9, it can be seen that the Bragg peak can be spread out into a plateau. The largest peak represents the highest energy of the beam. The lower peaks represent contributions from lower energy beams of different intensities, which when added together produce a plateau known as a 'spread out Bragg peak' (SOBP), meaning that tumours of appreciable thickness may be completely irradiated. It should be noted that the surface dose of a proton beam with a SOBP has a considerably higher value relative to that of a monoenergetic beam.

When it is not possible to vary the energy of the beam at the point of production, the energy of the beam incident on a patient may be varied by placing a rotating variable thickness wheel in the beam with open windows in it. When the beam passes through the open part of the wheel, the protons incident on the patient are at maximum energy. When they pass through the thickest part of the wheel, they lose energy and hence have a decreased range in tissue.

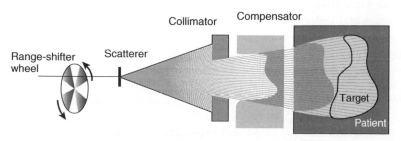

Fig. 4.10 Use of a passive scatterer and compensator to conform dose to distal tumour limit. Adapted from www.aapm.org/meetings/03SS/Presentations/Lomax.pdf with permission from Dr Tony Lomax.

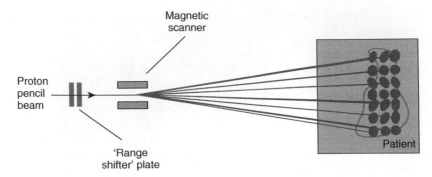

Fig. 4.11 Representation of spot scanning for protons beams. Adapted from www. aapm.org/meetings/03SS/Presentations/Lomax.pdf with permission from Dr Tony Lomax.

So—the beam which leaves the collimating system is a mixture of energies, the difference between the maximum and minimum energies being related to the thickness of the tumour to be treated. However, the beam modification is not yet complete. The rotating wheel will spread out the energy but do nothing to help shape the beam to conform to the tumour shape. The distal aspect of the tumour is usually closest to critical structures and we require a rapid fall dose fall off in this region. This is achieved by placing a static scatterer and compensator in the beam. The scatterer spreads the beam and the compensator modulates the lateral profile shape with depth to fit the shape of the target.

There is a major disadvantage to this method of production. While the dose can be conformed to the shape of the distal target volume, the spread of energies needed to cover the thickest part of the target cannot be changed laterally, meaning that if the tumour thickness changes laterally across the beam, there is generally an excess of normal tissue treated to a high dose on the proximal side of the target, as shown in Fig. 4.10.

With newer technologies offering better beam energy control, it is possible to treat a large tumour by scanning a narrow proton beam though a tumour, changing the energy according to the depth required. This is known as spot scanning or active scattering. By moving the beam throughout the volume, a more conformal dose can be delivered to the proximal side of the target volume as shown in Fig 4.11.

As with more conventional forms of photon radiotherapy, consideration needs to be given to patient and organ motion during treatment. Spot scanning offers the potential for more conformal doses but also presents the risk of missing the tumour due to organ or patient motion during treatment. While the more 'barn door' approach of passively scattered broad beams reduces conformality, the risk of geographic miss is reduced.

It has been demonstrated that beams produced by passive scattering have a higher neutron contamination than those generated by spot scanning and are hence considered less appropriate for the treatment of paediatric tumours.

Carbon ions are also of potential interest in cancer treatment. Being about 12 times heavier than a proton, they suffer even less lateral scattering and have a narrower lateral penumbra than protons. However, the dose deposited by carbon ions does not fall off as quickly as protons after the Bragg peak, which may be an disadvantage under certain circumstances.

4.6 Summary

In radiotherapy, damage to biological tissue occurs when energy disrupts biological molecules, in particular DNA. The energy is delivered by particles moving though the tissue.

In megavoltage photon therapy, these particles obtain their energy from moving photons, as described in Chapter 2. In particle therapy, be it electrons, protons, neutrons or carbon ions, it is the particles themselves that either damage the tissue molecules directly, or impart energy to electrons in the tissue.

The processes by which moving, energy laden, electrons interact with atoms in the tissue, irrespective of how they obtained that energy, are similar.

Chapter 5

Putting the IT in RT

N MacDougall and A Morgan

5.1 Introduction

In this chapter, we address the whole process of modern radiotherapy (RT) from introduction of patient demographic data into a Hospital Information System, through imaging associated with target definition, treatment planning and delivery to end of treatment and eventually follow-up as it works 'under the bonnet' in a fully electronic department. It also provides a simple set of references for all information technology (IT) concepts that occur in the book. If we can do this without losing you, then we'll be very happy!

We shall demystify frequently used terms such as DICOM, IP addressing and databases. From this point we should be able to demonstrate the possibilities and benefits of modern radiotherapy IT systems, but also the potential pitfalls and dangers that might befall the unwary.

5.2 Computers, who needs them?

Modern radiotherapy departments cannot function without IT support.

Even up to the early 1990s, much of the data used in the radiotherapy treatment planning and delivery processes had to be manually entered into the relevant computer systems. Most data produced by radiotherapy treatment planning systems was printed and then manually typed into the linear accelerator (linac). This was a time consuming process, was prone to errors and put a limit on the amount of data that could be used for a patient treatment. Things started to change in the mid 1990s with the introduction of multi-leaf collimators and the more widespread use of computerized tomography (CT) images in treatment planning. However, recent rapid increase in computing power has helped to transform radiotherapy, allowing large amounts of CT data to be accessed and highly complex mathematical models to be run to efficiently calculate the radiation doses to be delivered (IMRT, VMAT, IGRT).

5.2.1 Computers are everywhere

Imaging investigations—MRI, CT, PET, ultrasound etc.—all generate images through the use of computers, indeed none of these imaging techniques would be possible if it were not for the calculating power of modern computers. While generating an image is the start, we now have the problem of how to get these images off the CT scanner in a way that they can be read by another computer. In former times, the CT images would be printed onto a sheet of film. The person looking at them could easily tell that 'Mr Patient' was the patient's name and 'R1400400' was the hospital number. They could then look at all the pictures of the CT slices printed below and gain all the information they needed from the CT scans. However, we don't have to emphasize the advantage of not having to generate and store all that printed film. Anyone that has worked (or had to access historical data) in a paper and film based department will know of the limitations.

So, we want to do all this electronically, which makes accessing data easier, but setting everything up is a little bit more complex.

A computer is not as smart as a person (well, most people!) so it cannot reliably tell a name from another piece of information, such as a hospital number. Also, printed pictures contain a lot of information, which is takes a lot of electronic space to store. So, we have to get a bit clever to make this all work efficiently.

5.3 Data communication

One of the oldest problems of mankind is that of communication. The ability to concisely and accurately convey information is one that is highly valued. Just as one person talking to another must speak the same language, so computers talking to each other must talk in a similar 'language'. However, the analogy can be stretched further, even if two people speak the same language, different regional dialects can impede communication resulting in misunderstanding!

Now, keep that image in mind as we stretch the imagination to make computing a simpler place to understand.

Back in the dim and distant past (before DICOM—see below) all medical equipment talked in a local 'in house' style. To use the human analogy, each had their own language. So, for example, if we put a 'manufacturer X' MRI image onto a 'manufacturer Y' reviewing station nothing would be displayed unless one could write translator program to make the file readable.

5.3.1 DICOM

DICOM (Digital Imaging and Communication in Medicine) is a structured way of communicating electronic information. The DICOM file is a standard form (like an application form for a passport) that the sending computer fills in before attaching the image to the end. From the example in the previous paragraph if we introduce DICOM, manufacturer X and Y both agree that their imaging systems will write and read data in the same way: according to the DICOM standard. Now there should be no need for translation: the sending computer orders the information in exactly the way the receiving computer expects to see it. As the receiving computer, I know the first thing I will see is the patient's surname, so I know that whatever is written there is that value.

But, simply put, a DICOM image file starts with a 'header' which is just text that contains all the relevant details pertaining to the patient, hospital and imaging modality. At the end of the text comes the image (e.g. the CT slice), which is compressed to save space in tagged image file format (TIFF). If you imagine this like a word document with a patient report and a picture at the bottom, you'd be close!

If you're surprised, don't be. It really is this basic. And there is more than one type of DICOM, so it is not enough for a device to be 'DICOM compatible', it must be the 'right kind of DICOM' compatible. This is potentially a source of error that could affect patient treatment.

So, we've packed up the images in a nice parcel, now to get them to the radiotherapy department.

The correct name for the format of the images that go into the planning system is DICOM. All modern imaging techniques produce images that are in the DICOM format. All the image types that might go into a radiotherapy treatment planning system (such as CT, MR and PET) are in DICOM format. The headers for each image type will all be slightly different, but that doesn't matter. DICOM means that the planning system can understand the content of the headers and present the images to the user in the correct format.

There are five main sub formats of DICOM for radiotherapy (Table 5.2). To demonstrate their use, let's assume that we have a series of CT images on a planning system. Generally, the first thing that happens is that some outlines are drawn. These are usually the target volumes and critical organs. There are several mechanisms for performing contouring and the fine details are not important. What is important is that at the end of the process, we have a set of contours on which a plan can be designed. This set of contours can be stored in what is known as DICOM-RT-Structure set format. So the patient record now contains a set of DICOM images and a DICOM-RT-Structure set, which contains details on the number of contours drawn, number of points in each contour and their names. The next step is to put some beams on the plan.

This process generates another DICOM-RT file called the RT-Plan, which contains details of each treatment beam, such as its name, jaw settings, energy, monitor units etc.

Once the beams have been positioned, a dose calculation is usually done and as you might well have already guessed, this generates a DICOM-RT file called RT-Dose. This contains details of the dose calculation matrix geometry, dose volume histogram data etc.

At some point during the planning process, digitally reconstructed radiographs may be produced and there is a DICOM-RT format for these called RT-Image. Verification images taken using an electronic portal imaging device also generate images in RT-Image format.

When the plan is ready for treatment, some or all of these files may be sent to the linac control system. The RT-Plan is essential in this respect but some of the others are optional and the functionality available may be vendor specific. Every time the patient has treatment, the treatment parameters used are stored by a **record and verify system (R&V system)**. Again, at the end of treatment, the full treatment record can be stored as our fifth type of radiotherapy DICOM object: a RT-Treatment Record.

5.3.2 Record and verify systems

A R&V system is a database with various bits of software which allow most of the functions in radiotherapy to occur. R&V systems are named from the days when all they were used for was to record delivered treatments and ensure the same treatment got delivered from one day to the next. The computer systems used today are much more sophisticated, allowing all kinds of information to be added to the patient record, such as diagnosis, treatment plans, imaging. The list is almost endless.

Table 5.1 Simplified example of a DICOM file

Property name	Data
Study date	20110404
Modality	RT-Image
Patient name	Patient A
Patient ID	123456
Image Data	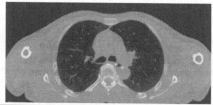

So where is this obsession with DICOM going? The major advantage of the adoption of an international standard such as DICOM is to increase the longevity of all patient data stored electronically. All equipment manufacturers provide mechanisms for backing up and archiving data. However in the pre-DICOM years all such data was stored in a manufacturer specific format, meaning that it couldn't be interpreted by another manufacturers' equipment. This might not seem much of a problem but, for example, when a department changed its planning system, physicists were usually left to destroy a pile of tapes or other media containing data from the old system which couldn't be used on the new one! Now, if all manufacturers implement DICOM uniformly, this scenario will be unlikely to arise as the data from the old system will be readable by the new one (or by another manufacturer's system in different hospital). This means that if a patient undergoes treatment in one centre, then moves house and requires treatment later in another centre, the old record may be accessible, eliminating some of the guesswork that currently occurs in such circumstances!

Table 5.2 Radiotherapy DICOM objects

DICOM-RT object	Main property	Example contents
RT-structure	Patient anatomical information.	PTV, OAR, other contours.
RT-Plan	Instructions to the linac for patient treatment.	Treatment beam details e.g. gantry, collimator and couch angles; jaw and MLC positions.
RT-Image	Radiotherapy image storage/transfer.	Simulator, portal image.
RT-Dose	Dose distribution data.	Patient dose distributions (in 3D), Dose volume histograms.
RT-Treatment Record	Details of treatment delivered to patient.	Date and time of treatment, MU delivered, actual linac settings.

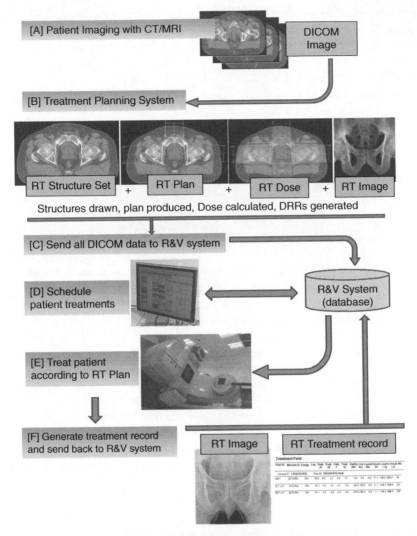

Fig. 5.1 The ABC of DICOM. DICOM-RT and the radiotherapy pathway. This figure is reproduced in colour in the colour plate section.

5.4 **Networking**

A single computer sitting by itself is of some use but if a computer can communicate with other computers, its potential use increases many fold. A group of computers that communicate with each other is known as a **network**. There isn't a minimum number of computers needed to define a network—there may be just five or five hundred (the internet is just a big network.). The important matter is that they can communicate with each other quickly and efficiently, getting data to where it needs to be in a timely manner. In a hospital environment, computers are generally connected by a physical network— usually a cable. The methods of connecting computers with cables are too numerous to describe but one of the simplest ways is to connect computers is using a device known as a **hub**. A hub is simply a box which cables from all computers go to and which enables computers to transfer data between each other (analogous to a crossroad).

5.4.1 IP addressing

In order to transfer data between computers, they need to know each others identities. While the humans operating them might know them as 'The plannning system' or 'Linac 1 PC', the computers use a different notation known as an IP (Internet Protocol) address. This is usually a twelve digit number, made up of four groups of three numbers separated by full stops. A typical example might be 105.234.100.185. Each group of three numbers must be within the range 0–255. This gives a possible 4.3 billion IP addresses. On any given network each computer must have a unique IP address or things get very confused. It's a bit like giving critical instructions to two people called Bob, e.g. '"Bob I'd like you to carry out activity X" and "Bob, I'd like you to carry out activity Y"—clear? good'. In this scenario, neither Bob knows what they're meant to be doing and so will probably do nothing, or both try to do it and conflict with each other. It is the responsibility of the network manager to make sure that each system is allocated a unique IP address.

5.4.2 Data storage

A device that's generally also associated with a network is a **server**. It is possible for users to store data on each computer they use. However, there are very good reasons for not doing so:

1) If users then move to another computer, they may not be able to access the data they were working on previously,

2) If the computer malfunctions or breaks down, data being worked on may be lost,

3) If the computer gets stolen, data may be lost and if this data identifies patient related information, there will be mountains of paperwork to fill in and jobs may be lost!

Therefore it is common practice to use a data server. A server is a computer a bit like a bigger version of your desktop PC but with some extra parts and features to enable it to continue working if one individual part of it fails. It also runs software which allows many users to access it simultaneously. So a server gets around the above problems, because:

a) All users can access the server,

b) The server creates several copies of the data stored on it so if one part fails, backup copies already exist,

c) The server is located in a locked cabinet in a restricted access room so the chances of it being stolen are minimized.

There may well be more than one server on a network. Servers are designed to attempt to protect data integrity. Most PCs have one hard disc where data is stored. If that hard disc fails, all data stored on it is likely to be lost—or at least corrupted in some way. Servers use what is known as a RAID architecture (Redundant Array of Independent Discs). All data stored on them is copied and shared between several independent hard discs (minimum of 2 discs). The server controller knows where all the data is stored—if one disc fails, it can easily be replaced and the data that was

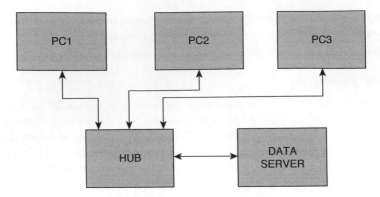

Fig. 5.2 Generic representation of computer network.

stored on it is restored from other discs in the RAID. Of course, this is absolutely no use if someone steals the complete server—which is why such devices are usually stored in remote, secure rooms with restricted access.

5.4.3 Simple network layout

Having introduced the main elements of a network, let's look at how a typical small network may be represented. In a typical radiotherapy department, this may equate to that in Figure 5.3.

So, putting it simply, patients will have a planning CT scan. The CT scans will be sent to the treatment planning system where a plan will be produced. The treatment plan will be sent to the R&V system where it forms the basis for the patients' radiotherapy treatment record. From here, the necessary linac parameters and dose data is transferred to the treatment unit computer on a daily basis for treatment. Any treatment amendments or notes can be added to the record and stored back on the R&V system for recall the following day.

These diagrams represent very simple networks and in larger departments, such diagrams will get very complex once we've added in other equipment such as portal imaging devices, cone beam CT units, other simulators (both CT and conventional),

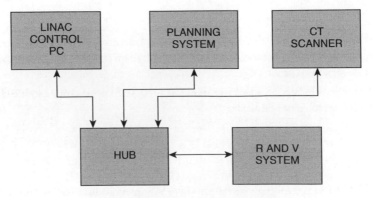

Fig. 5.3 Generic representation of simple radiotherapy network.

MR and PET scanners. However, the basic principals will remain the same as described earlier. Each computer system on the radiotherapy network needs to be assigned an IP address to enable it to communicate with other computers on the network.

5.4.3 Network security

One prospect that terrifies IT and radiotherapy physics departments alike is the potential for introducing computer viruses onto a network. Most computer systems use Windows software and are potential targets for malicious software writers. Viruses can be quite easily and usually quite innocently introduced into a department. Staff often work at home on PCs which may be infected and transfer the virus to a network via a memory stick or similar device. Once a file is opened on the network, it spreads to other PCs. Most viruses can be easily dealt with by virus checking software which warns users of any threat. However, some systems cannot run anti-virus (AV) software. For example, computers that are provided with linacs and treatment planning system manufacturers are considered to be 'medical devices'. These were developed and tested by the suppliers without any AV software present on them as AV software may cause the systems to malfunction. There are many different AV software packages and it is not possible to test them all. Therefore medical devices must be protected and this is often by use of a **firewall**. A firewall is a box (a bit like a hub) that sits on the network between the medical device computer and the rest of the network, like a bouncer on a night club door! It has rules about what it will and will not allow to pass through it, so in theory it will allow computers to do what is expected of them, but block the sorts of activities which viruses might initiate. Other methods of reducing the risk of virus introduction include disabling the use of CDs and USB ports for memory sticks and blocking access to the internet, or if internet access is allowed, limiting it to certain websites from which viruses may be imported. Many radiotherapy departments have experienced computer virus outbreaks and the impact can be devastating, occasionally stopping patient treatment for days.

5.5 Patient safety

Perhaps one of the greatest benefits of computing in radiotherapy is the reduction of transcription errors (by only storing information in one place). The radiotherapy pre-treatment process involves many different groups of staff adding new information and modifying existing information on the patient's treatment plan. If one is to carry this out manually, then an accurate patient treatment relies on all staff concerned transcribing important information with no errors from one piece of paper to another. This can be particularly troublesome with numbers and decimal places!

If one can generate the information for patient treatment on the planning system and then save it on a server that all other staff, and the linac, can access, then there is no need for anyone to copy any important data. But how do we store all this information electronically in a way that is safe and accurate? We use software called a **database**.

5.5.1 Databases

Don't give up at this point just because of the title! Databases are quite simple beasts if we don't get too technical. The main aim here is to store information in an orderly manner (so you can find it again!) with no repetition. The best way to accurately keep a record of a piece of information and be able to update it is if there is only one copy of it. At a basic level, a database is a way of storing large amounts of related data in the simplest way possible. There is lots of complex maths behind this, but let's not go there. If we consider the information required to treat a patient with radiotherapy, this should all be self-explanatory.

In a patient's paper notes (if you still have them!) every piece of paper should have a patient ID sticker on it. This will contain information unique to the patient: first name, surname, ID number, address. This is a sensible approach, for if an important piece of paper (e.g. prescription card) is dropped from the patient folder it will be instantly identifiable and easily reunited with the patient notes. However, what if we discover that the patient's ID number on the ID sticker is incorrect (due to a data entry problem)? There may be tens (or hundreds) of stickers on the notes, treatment plan, images etc. and we'll have to manually correct every one of them!

However, if this information is stored in a database, then it can contain all the information in the notes, but organized to store each unique piece of information in one place only. As such, in its storage form, it would not make much sense. However, the database is a bit clever.

If we consider the patient's ID sticker. We would store the information to generate unique patient IDs using the following identifiers:.

- ◆ Unique identifiers (that cannot be changed) e.g. name, date of birth, NHS number and hospital number. We store this information together in a 'demographic table'.

- ◆ Identifiers that can be changed by the patient e.g. telephone number and address. We store these in an 'address table'.

We then link these together e.g. the demographic table can have many address tables.

We are saying that whilst a patient can only have one copy of the demographic table (therefore only one DoB, name etc.) they can have many copies of the 'address

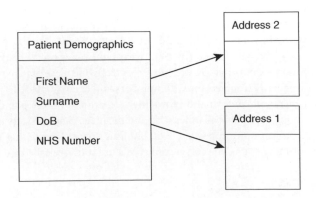

Fig. 5.4 Simple example of a patient with two addresses.

table'. Therefore, more than one address, phone number etc. These two tables are joined together and the join knows this rule. This rule is a relation, which makes a **relational database** (albeit a very simple one). This concept can be easily expanded to include the course of treatment (patients can have many courses, unfortunately), each course can have many phases etc.

By storing the patient's data in this way we only have to store (or change) information in one place. So, if we enter Marjory Bloggs particulars into this database and you are about to authorize a completed electronic treatment plan, but, oops, you notice we've spelt her name wrong! You can change it in one place and all the multiple plans, images etc. will now be linked to this new name you have entered. This is an improvement on changing thirty ID labels in the paper example. Press print and any paper reports will come out correctly too.

The possibilities, from this simple start, are almost endless. At a press of a button you can retrospectively gather all kinds of information from clinical trial data to workload figures. However, it is important to stress that if you don't put the information in at the start, then it won't be there later.

The cynical among you may be thinking, so very good, you have all the information in one place and once it is stored the patient's treatment will always be the same day after day. Well, what if it is stored incorrectly? This is a very real problem, computers can ensure consistency, but it is the job of clinical staff to ensure what is stored there is correct from the start. To this end there are many checking procedures that occur in a patient's route to treatment in radiotherapy. This is the case if the treatment is paper based or electronic.

5.6 **Summary**

Reading this chapter won't make you an IT expert, but it should remove some of the fear of dealing with electronic issues in a radiotherapy department. The central message is that whilst IT has enabled radiotherapy to be immeasurably more sophisticated it is not always as turnkey as one would desire.

Chapter 6

Principles of imaging for radiotherapy

V Khoo and N Van As

6.1 Introduction

The developments in radiation delivery in the last decade have been dramatic. We can now accurately deliver image guided and/or adaptive radiotherapy. However all the advances in image guided radiotherapy (IGRT) rely on accurate definition of target volumes, and hence imaging is crucial at all stages in the radiotherapy process.

A big change in design of treatment fields occurred when plain X-rays with bony anatomical landmarks were used instead of visual or physical palpation of abnormal masses. The revolution in radiotherapy began when cross sectional imaging was used to identify target volumes and to design treatment fields. The ongoing challenge in radiotherapy is to improve the therapeutic ratio by enhancing dose delivery to tumours whilst sparing normal tissue. Advances in imaging technology and medical equipment have brought many radiotherapy developments from improved staging to more reliable treatment delivery and verification.

It is appropriate to define what is meant by radiotherapy treatment planning (RTP) as this process can include the initial step of cancer staging to designing the radiotherapy plan both of which are outside the scope of this chapter. For the context of this chapter, RTP is defined as the application of imaging for target volume delineation. This chapter will review and summarize the rationale of imaging methods for modern radiotherapy, the utility of multimodality imaging in RTP, the application of functional/biological imaging for RTP, innovations in technological imaging for IGRT and future directions. In addition we will address how imaging is now incorporated into radiotherapy treatment delivery (RTD).

6.2 Types of imaging

6.2.1 X-rays

Plain X-rays have been traditionally used in radiotherapy. The radiotherapy simulator is a diagnostic X-ray machine, mounted on a gantry similar to a linear accelerator (linac) with a couch to set up the patient in the treatment position. X-ray images are obtained for the same beam orientation on which field borders are set. The technique is limited by the lack of soft tissue information and has been largely superseded by 'virtual simulation' using a computerized tomography (CT) scanner.

6.2.2 **Computerized Tomography**

The general availability of CT dramatically changed RTP by allowing the tumour not only to be accurately defined in 3D but in relation to its anatomical surroundings. This is a prerequisite for 3D beam shaping and forms the basis for conformal radiotherapy (CFRT). The target volume can now be fully individualized with 'conformal avoidance' of dose limiting normal organs. Following delineation of target volumes on contiguous axial images, these axial slices are then stacked to create a reconstructed 3D volume where the field portals are designed using beams eye view (BEV) facilities. The BEV is the view of a virtual observer as seen along the projection of the treatment field from the source of radiation within the linac. This visual perspective allows a projection of the planned volume in relation to the field borders and allows for maximal avoidance of adjacent organs at risk (OAR).

6.2.3 **Magnetic resonance imaging (MRI)**

MRI utilizes the natural magnetization induced in the body when placed in a strong magnetic field. Each tissues' protons give off a very small magnetic field created by the spin of the protons. Prior to entering the magnet, the magnetic field of all protons will be in random directions. When the body is placed in the magnetic field the protons will align along the gradient created by the field: this direction is referred to as B_0. Whilst the protons are aligned along B_0 it is not possible to detect any signal. Signal is created when the protons are knocked out of alignment, which is achieved by applying a radio frequency pulse wave across the magnetic field. When the protons are knocked out of plane they give off a signal detected by the scanner. By varying both the magnetic gradient and the strength and timing of the radio frequency pulses different signal characteristics can be created for the same tissue thus improving its evaluation.

6.2.3.1 **MRI rationale for RTP**

MRI is being increasingly used in oncology for staging, assessing tumour response and evaluating disease recurrence. Indeed, MRI has replaced CT as the diagnostic imaging modality of choice for many body regions. The advantages and disadvantages of MRI for RTP are outlined in Table 6.1 and include greater functionality in characterizing soft tissues with similar X-ray attenuation properties for better tissue discrimination of the tumour and its boundaries thus improving reliability for target volume delineation. Other benefits include more reliable and consistent target definition to reduce inter-observer and intra-observer variability.

6.2.3.2 **Some applications of MRI for RTP**

For central nervous system (CNS) radiotherapy, MRI is used extensively for staging and radiotherapy planning. Quantitative improvements of up to 80% in target coverage have been reported when MRI is added to CT planning. MRI and CT may also provide complementary information. For base of skull meningiomas where CT X-ray attenuation from large skull bones may obscure soft tissue detail, MRI images provided better visualization of tumour extent especially for disease that crept along the skull bones. However, CT showed tumour related bony erosion unavailable on MRI demonstrating complementary information that was useful in optimizing target volumes. The use of

CT-MRI co-registration for CNS RTP may now be considered as standard practice and automated and atlas based segmentation algorithms may aid this process.

Determining the extent of tumour infiltration in the head and neck region can be difficult because of the complex anatomy at this site. In nasopharyngeal cancers, in a study of over 250 patients, CT imaging missed up to 40% of intracranial infiltration detected by MRI. MRI can aid RTP in this region by defining the extent of tumour infiltration at the following sites:

- The extent of peri-neural infiltration and intracranial extension e.g. nasopharyngeal tumours;
- Tumour infiltration of soft tissue structures and tissue planes such as the pterygoids and tongue;
- Longitudinal tumour infiltration along the upper aero-digestive tract and adjacent fascial planes e.g. pre-vertebral fascia.

Specific MRI segmentation algorithms based on the contrast enhancement ratio of T1-weighted and signal intensity of T2-weighted images are being developed to aid the delineation process for RTP.

In rectal radiotherapy, CT planning has provided 3D visualization of rectal tumours, the mesorectum and nodal regions compared to traditional planning using plain X-rays and rectal contrast. Limitations in CT definition of rectal tumours may occur due to poor contrast between faeces and tumour, partial volume effects due to the curves/valves of Houston in the rectum and imaging of the horizontal sigmoid. Anal canal/sphincter infiltration in low rectal cancers can be difficult to assess unless there is an obvious mass effect. This can be improved by MRI by better definition of rectal wall invasion, spread into the mesorectum and any longitudinal tumour spread along the rectum. A planning study comparing CT and MRI to determine CTV showed that CT consistently overestimates the extent of the tumour compared to MRI. By improving the accuracy and reliability of the tumour extent, strategies of anal sphincter sparing, tumour boosting and dose escalation with or without concurrent chemotherapy may be initiated.

In prostate RTP studies comparing co-registered CT-MRI scans have reported that CT defined prostate volumes are 27–33% larger than MRI volumes suggesting an overestimation with CT that is due to CT uncertainties. In prostate radiotherapy, better delineation of erectile tissues and rectum by MRI can permit dose sparing of these structures by IMRT and may improve patient outcomes. MRI can also be useful where internal anatomy has been substantially altered due to previous extensive surgery such as abdominal-perineal resections and MR can be of benefit in target definition in the presence of bilateral hip replacements (Fig. 6.1). Improved delineation of prostate and seminal vesicles can also reduce intra-observer and intra-observer variation and minimizing planning variance within departments and clinical trials.

Multimodality CT-MRI imaging is also useful in brachytherapy. Investigators have reported improved target volume definition, reduced observer variability, increased confidence in needle placement compared to ultrasound or CT. MRI is useful for post-implant dosimetry as delineating the prostate gland with CT is maybe difficult following implantation due to seed induced artifact and the intrinsic poor tissue contrast of CT.

Table 6.1 Advantages and disadvantages of MRI for RTP

Features	Advantages	Disadvantages
Patient	Non or minimally invasive procedure. Few patient risks. No radiation associated with imaging. This may be advantageous to paediatric patients, and pregnant women. This may be a useful for follow-up scanning.	Claustrophobia due to the smaller patient bore. Contraindicated in patients with loose metal foreign bodies within the body particularly the orbits, or pacemakers.
Imaging	Increased imaging parameters for more imaging flexibility. Superior soft tissue imaging with excellent spatial resolution to provide better visualization for the following: ◆ Determining the tumour/GTV extent and degree of tumour infiltration ◆ Understanding the surgical bed or altered anatomy secondary to surgery. ◆ Distinguishing between post-treatment fibrosis or tumour recurrence. ◆ Improved definition of normal soft tissue structures and tissue planes. ◆ Avoidance of image artefact from metal prosthesis and large bony regions. ◆ True multiplanar capability to image in any oblique plane and reduction of the 'partial volume' imaging effect. ◆ Increased accuracy, reliability and consistency of target definition to reduce both inter- and intra-observer variability. ◆ Providing functional and biological information for functional avoidance or biological targeting. Ultra-fast volumetric and cine mode acquisitions to assess temporal-spatial variations in target positioning or deformation. Can be registered with CT information for use in RTP systems.	MR image distortion: ◆ Systems, ◆ Object induced distortions. Lack of electron density information for dosimetry and needs additional steps to permit dose calculations. Lack of cortical bone information to create digitally reconstructed radiographs. (DRR) in radiotherapy May have longer scan times than CT with more potential for motion artefact. Need for specific training to comprehend and understand MR images for RTP use. RTP systems can only import transverse MR images and cannot take full advantage of sagittal and coronal in-plane MR images. Most immobilization devices used in radiotherapy may not be MR compatible.
Contrast agents	New contrast agents (i.e. USPIO) to define nodal status. Less incidence of allergic reactions to gadolinium than iodine-based contrast agents.	
Machine	New bore flange openings to lessen patient claustrophobia. Open MR systems for easier patient access, tolerance and positioning for radiotherapy.	Not as readily available and accessible. Smaller bore than CT (52 cm vs 82–85 cm). Curved table top.

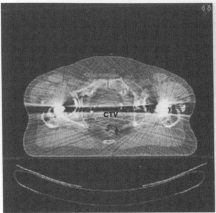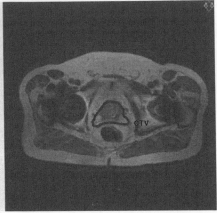

Fig. 6.1 The use of CT-MR image registration to define target volumes in pelvic radiotherapy in the presence of bilateral hip replacements.
Charnley, N., *et al.* (2005). The use of CT-MR image registration to define target volumes in pelvic radiotherapy in the presence of bilateral hip replacements. *British Journal of Radiology* **78**: 634–636. This figure is reproduced in colour in the colour plate section.

Co-registered MR-CT images have also been useful in other brachytherapy sites, such as for head and neck cancers, sarcomas, and gynaecology. The use of MRI for cervical cancer brachytherapy has been endorsed by the Gynaecological GEC-ESTRO Working Group.

6.3 Dynamic contrast enhanced MR (DCE-MRI)

DCE-MRI is a technique which produces dynamic data when rapidly repeated sequential imaging sequences are performed during injection of intravenous contrast media. Many different parameters can be evaluated with DCE-MRI. The most straightforward approach to the quantification of enhancement is to directly compare the signal intensity curves with regions of interest. This is achieved by using measurement of the time taken for the tumour tissue to attain 90% of its subsequent maximal enhancement (T90). The maximum rate of the change of enhancement can also be measured. Some issues of interpretation reliability using these signal intensity based methods include the non-linear relationship between contrast enhancement and signal intensity. Despite these limitations, they remain the most common way of assessing DCE-MRI in the clinical setting.

6.3.1 MR spectroscopy (MRS)

MR spectroscopy (MRS) utilizes magnetic field spectra from endogenous chemicals present in tissue. As an example, in normal prostate the concentration of citrate is high and choline is low. Choline is a marker of cell membranes and is elevated in prostate cancer. The difference in concentration of metabolites is probably due to enhanced phospholipid cell membrane turnover associated with tumour cell proliferation.

MRS currently has limited applications in oncology. It is time consuming and requires very specific physics expertise.

6.3.2 Diffusion weighted MRI (DW-MRI)

DW-MRI provides information associated with the molecular movement of water in biological tissues. The degree of motion measured by DW-MRI relates to the mean path length travelled by hydrogen nuclei in the body within an observation period. Longer path lengths are associated with higher apparent diffusion coefficients (ADCs). The ADC is calculated by acquiring two or more DW-MRI images, identical in every aspect other than their sensitivity to diffusion. The extent to which a DW sequence is sensitive to diffusion is described by its b-value. Higher b-values are more sensitized to the detection of motion. The 3D diffusion of water within the body is not truly random because natural barriers impede diffusion. Within tissues these barriers are cell membranes. In the case of highly cellular tissue, an increased number of cell membranes may significantly impede the motion of extracellular water leading to a lower path length and hence lower ADC value. Consequently, parameters derived from DW-MRI indirectly provide information regarding the cellularity of a given tissue.

These MRI techniques may be used to select regions within the tumour or organ for individualized targeting or dose boosting using IMRT with IGRT. However there remains considerable work to validate the concept and to define its feasibility and clinical effectiveness.

6.3.3 MRI issues for RTP

There are several issues to consider when using MRI data for RTP (Table 6.1). The two main features are correcting MR related distortions and overcoming the lack of electron density information needed for assessing tissue inhomogeneities and dosimetry.

MR image distortion is important as inaccuracies resulting from this will limit the value of CFRT or IMRT and could be propogated throughout radiotherapy. MR image distortions can be grouped into two main categories; system related and object induced. To use MR images for RTP, these image distortions must be evaluated, quantified, minimized, and/or corrected. Using coordinated phantoms with preset marker based array coordinates, mapping of system based and object inducted distortions can be charted and distortional shifts of up to 5mm can be corrected (Fig. 6.2). Site specific phantoms such as with air cavities can be used to assess the use of MRI for lung RTP. It is important to ensure that the same MR scanner and imaging sequence used for mapping is also used to image the patients. Once MR image distortions have been quantified and corrected, image co-registration will allow the MRI information to be transposed on CT images and used for RTP and dosimetry.

6.4 Positron emissions spectrum for RTP

Positron emissions spectrum (PET) scanning can improve staging by detecting unsuspected metastatic disease and avoid treatment of these patients with radical therapy. PET can also be used to assess radiotherapy response, locate disease not readily identified

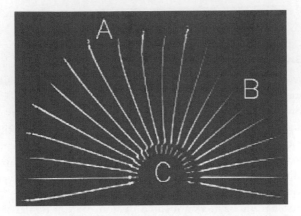

Fig. 6.2 An illustration of various forms of distortion in MR imaging using a phantom consisting of a coplanar array of water filled tubes embedded within in a circular solid plastic (PMMA) block. System distortion effects are seen in the apparent curvature of the tubes at A and their disappearance at B, which was due to warping distortion of the imaging plane. Magnetic susceptibility differences due to the presence of the plastic support block at C give rise to object induced distortions in the form of discontinuities at the point where each tube enters the support block.

by morphological imaging for target volume delineation and differentiate tumour from necrosis or benign processes such as atelectasis.

PET provides physiological information for normal tissues and disease regions. The most commonly used PET tracer is 2-[F-18]fluro-2-deoxy-D-glucose (^{18}F-FDG) and relates glucose metabolism via a combination of mechanisms to the increased metabolic activity seen in cancer tissue. Newer PET tracers can assess other aspects of tumour metabolism such as hypoxia and proliferation and these may identify tumour regions for selective dose boosts.

FDG-PET has limited utility in the CNS regions because of the normally high glucose uptake in the brain making it difficult to differentiate normal brain tissue from tumour. More specific PET tracers, such as radiolabelled amino acids; methionine (MET), alanine and tyrosine are being used assuming that cellular turnover in much greater in tumours than in normal brain tissue. Whilst MET-PET has been reported to have a sensitivity of 87% and specificity of 89% for the detection of tumour tissue relative to normal brain tissue, there are only a few reports of MET-PET for target volume delineation. These small studies suggest that MET-PET can aid target volume delineation in both low and high grade gliomas as well as meningiomas by more reliably distinguishing postoperative changes from the residual tumour and reducing inter-observer variation.

The main contributions of PET for lung RTP are to identify involved mediastinal nodes and to better distinguish between tumour and lung atelectasis. There is considerable clinical uncertainty in determining the boundaries between tumour and atelectasis using CT. This can be improved by PET and can lead to smaller treatment volume with fewer side-effects. PET based planning can also reduce inter-observer variability with this effect being most pronounced for target regions in the mediastinum and in excluding atelectasis. This benefit of PET co-registered with CT has been recommended by the American Radiation Therapy Oncology Group (RTOG) for all dose escalation trials in lung cancer.

PET can change RTP for head and neck radiotherapy in some cases. A small but robust study by Daisne revealed that both CT and MRI substantially overestimated the tumour volume with the smallest discrepancy between FDG-PET and histopathology specimens. The salient feature was that none of the 3 imaging modalities accurately defined the microscopic disease extent. FEG-PET has also improved inter-observer concordance during head and neck RTP. Planning studies using the reduced target volumes from PET-CT have reported improved dosimetry with more normal tissues sparing.

For gastrointestinal tumours, especially colorectal cancers, the use of PET is mainly focused on staging, assessing response and recurrence. In oesophageal cancers, one study of 30 cases outlined that 47% of cases had more abnormal lymphadenopathy with FDG-PET than defined by transoesophageal ultrasound and this is likely to change the CTV.

Studies have shown using FDG-PET for rectal RTP altered GTV delineation in 15–67% cases and in anal RTP FDG-PET altered target volumes in 33–57% of cases. There is no consistent trend in whether treatment volumes are increased or decreased using FDG-PET and further studies are needed to define its utility for gastrointestinal RTP.

The urinary excretion of FDG in the pelvis limits its use for RTP especially in prostate RTP. However, choline-PET appears promising as choline is associated as marker of cellular membrane synthesis with high uptake in prostate cancer. Identification of choline-PET avid regions may allow targeting of these intraprostatic regions for dose boosts using IMRT.

PET is used extensively for lymphomas but it has not greatly impacted on the design of its treatment fields as they are generally regional treatment volumes. Furthermore, its role has been questioned as it has been reported that a negative PET scan following therapy does not exclude residual microscopic disease.

6.5 Imaging for treatment delivery

We will describe the types of imaging techniques available to guide the delivery of radiotherapy. There are essentially two broad categories, imaging methods that are added to conventional linear accelerators, and purpose built machines the integrate imaging into the radiotherapy machine.

Online strategies can be grouped into those that provide bony landmark identification or radio-opaque fiducial marker localization such as 'stereoscopic' kV fluoroscopy or those that provide cross sectional information such as the combination of a CT scanner in a therapy room or integrated in the linac to provide MV-CT images, cone beam CT on a linac.

6.5.1 IGRT with a linear accelerator

Historically day to day set up was verified by portal imaging using the treatment beam to generate MV images. However using portal imaging is time consuming, provides

Fig. 6.3 Amorphous silicon EPID.

relatively poor quality images and daily imaging would not be realistic in busy departments. The necessity for an online system spurred the development of Electronic portal imaging devices (EPID). Electronic portal imaging is the term used for a number of different methods of generating online images. A detailed description of all the EPID's available is beyond the scope of this chapter, but examples include a CCD video camera, liquid ion chamber and amorphous silicon flat panel detectors (Fig. 6.3) to create a digital image. For a detailed review of EPID's refer to 'Clinical Use of Electronic Portal Imaging: Report of AAPM Radiation Therapy Committee Task Group 58. EPIDs are a planar device and therefore require multiple gantry positions to provide 3D data. By placing another set of imaging detectors perpendicular to the treatment beam, volumetric data can be obtained. A cone beam image results from a single rotation of the system. Cone beam CT acquires many projections over the entire volume of interest in each projection. Using reconstruction strategies the 2D projections are reconstructed into a 3D volume analogous to the CT planning dataset.

6.5.2 **3D CT based image guidance**

3D images for treatment delivery can be obtained with KV or MV imaging, and either a cone beam or a fan beam. There are essentially four solutions, which are commercially available. Each technique will be described.

1. Kilovoltage fan beam CT (CT on rails)

2. Kilovoltage cone beam CT

3. Megavoltage cone beam

4. Megavoltage fan beam (Tomotherapy)

6.5.3 **Kilovoltage fan beam CT (CT on rails)**

CT on rails requires a bunker large enough to house a linac and a CT gantry. The patient is set up on the linac and then the couch is moved to an imaging position where the CT on rails slides over the treatment couch and obtains diagnostic quality kV CT images which are co-registered with the planning CT. The image quality is excellent, but the disadvantages are the that errors can occur due the movement of the treatment table between treatment set up and scan, and due to the design no information is available on intra-fraction patient or organ motion.

6.5.4 **Kilovoltage cone beam CT**

Cone beam CT (Fig. 6.4) is achieved using a diagnostic X-ray source and a flat panel detector integrated with a linac. One major advantage of this system compared to 'CT on rails' is that the patient is not moved between treatment set up and scan. The image quality is inferior, and the scan takes longer to acquire. Motion and breathing artefact further compromise the image quality. Despite the relatively poor image quality, the system is primarily designed to assess inter-fraction set up errors, and many authors have reported a wide range of uses for cone beam CT.

6.5.5 **Megavoltage cone beam**

Megavoltage cone beam systems were initially developed in the 1980's and were based on a single slice tomogram generated through one gantry rotation. As discussed earlier with EPID's the MV quality of images is poorer than kV CT, but there are fewer artefacts produced by prostheses, and accurate identification of fiducial markers. Alignment of the target is simple as the treatment beam is used to acquire the image. A major problem

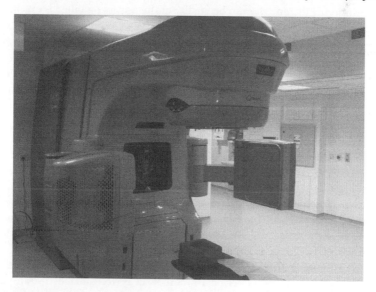

Fig. 6.4 Linac with kV imaging system deployed, which can be used for cone beam CT acquision.

with MV-CT is the relatively high dose received by the target whilst acquiring an image. Others have proposed low dose solutions showing that using a linac with low dose rate and an EPID clinically useful images can be obtained.

6.5.6 Megavoltage fan beam (Tomotherapy)

Tomotherapy is a technique whereby radiation is delivered by a beam rotating around a patient. Basically it is a CT scanner where the diagnostic X-ray tube has been replaced with a linac and the collimating jaws replaced with a binary collimator consisting of small high-density metal leaves. CT image acquisition, using a lower energy than for treatment, is acquired with all leaves open prior to treatment. The machines essentially intergrate IGRT and IMRT. The system incorporates a rapid auto-matching system, which enables daily online positional correction prior to treatment. This allows for correction of both random and systematic errors. The couch moves through the cone as the gantry rotates delivering radiation in a helical manner. The onboard CT can be fused with the planning CT for automated target localization prior to treatment. Phantom studies have accuracy of 1mm or 1 degree of rotation. The set up correction can be applied either by moving the patient or altering the IMRT delivery. Alternatively the CT detectors can be operated during treatment and compared with the expected signal.

6.6 Online stereoscopic kV fluoroscopy

Several systems have been developed to enable the localization of internally implanted markers or fiducials using kV imaging systems that are housed on the gantry of the linac or within the therapy room. The quality of the images obtained is excellent and permit the identification of fiducials down to the size of 3mm in length. An example of this system uses a linac synchronized with a fluoroscopic real time tumour tracking system which has four sets of diagnostic X-ray television systems, each adjusted such that the central axis of the individual systems would cross at the isocentre of the linac. With four imaging systems, continuous stereoscopic monitoring of the fiducial markers can be undertaken regardless of the positioning of the linac treatment head. Using a moving object recognition software system, this system has been reported to have an accuracy of +/− 1 mm in determining the position of the implanted marker every 0.033 seconds during radiotherapy. The Exac Trac system developed by Brain Lab is based on a similar concept. There are two kV X-ray units recessed in to the floor of the linac room with two ceiling mounted flat panel detectors. An integrated optical infra-red tracking system monitors the patient's position during treatment. The patient's position can be altered remotely by moving the couch. The system calculates the required patient shift and the remote couch aligns to the correct isocentre. Studies have shown the system is accurate to 0.5mm and can be used to deliver frameless radiosurgery for brain tumours. An alternative to kV marker localization is the Calypso system. This system employs fiducial markers that act as electromagnetic transponders and a detector. The transponders and detectors work like a 'GPS' system, with real time coordinates of the target available. An advantage of the system is that it does not require ionizing radiation, but as it relies an electromagnetic field it is not compatible with MRI. The calypso system has been predominantly used for prostate cancer, where planning studies have shown the potential to reduce the dose to the rectum.

Another example of an in-room kV monitoring system is the Cyberknife (Accuray Inc, Sunnyvale, CA, USA). The Cyberknife is a 6 MV linear accelerator mounted on a computer controlled robotic arm. Cyberknife is a system for delivering for delivering highly conformal, image guided, adaptive, and high dose radiotherapy. The major advantages of the system are that it continuously tracks and corrects for movement of both the patient and the tumour and can deliver radiotherapy with 0.5mm accuracy. Cyberknife uses a number of integrated technologies. Firstly, a lightweight linear accelerator is mounted on a precisely controlled robot. The robotic mounting allows very fast repositioning of the radiation source, which enables the system to deliver treatment from multiple different directions. It is equipped with an orthogonal pair of diagnostic quality digital X-ray imaging devices that monitors the position of implanted markers. The X-ray camera images are compared to a library of computer generated images of the patient anatomy. Digitally Reconstructed Radiographs (DRR's) and a computer algorithm determine what motion corrections have to be given to the robot because of patient movement. This imaging system allows the CyberKnife to deliver radiation with an accuracy of 0.5mm and a tracking error of <1mm to the position of the radio-opaque implanted markers. At present most soft tissue tumours require the insertion of fiducial markers, but fiducial free systems are being tested.

6.7 Conclusion

Imaging is now at the heart of all aspects of radiotherapy planning and treatment delivery. Improvements in technology offer the potential to enhance the therapeutic ratio, with the potential for both improved oncological outcomes and reduced patient morbidity. It is crucial that as we embrace these technological advancements, we test the techniques in well-designed and ideally randomized controlled trials. Rigorous quality assurance is essential, and we should never lose sight of the potential for very 'precisely delivered' geographical misses.

Chapter 7

Radiation dosimetry

T Greener and J Byrne

7.1 **Introduction**

Accurate determination of dose is crucial to the success of radiotherapy. Errors in dose determination can cause failure of control or unacceptable normal tissue damage. Accurate determination is vital in clinical trials where the assumption is made that the observed response has been caused by delivery of a particular dose. Dose measurement is difficult but for accurate radiotherapy it needs to be better than 7% accurate and preferably better than 5% accurate.

Radiation emanates from a source, travels a distance and then interacts with the material through which it is travelling resulting in deposition of dose. When a dose measurement is made, the value obtained from the device is an indication of the energy deposited in the device itself and not to the patient or material (e.g. water or other phantom) in which the device sits. It is important to understand how the output of the measuring device (detector) is converted to the dose in the medium in which the detector sits.

7.1.1 **Absorbed dose**

Dosimetry defines a numerical relationship between ionizing radiation and the effect that it produces. It is logical to assume that this effect will vary according to the amount of energy deposited within a material of a given mass. Likewise, for the same amount of energy deposited the effect would differ if the energy were deposited within a large rather than small mass of material. The definition of absorbed dose is simply the expression of these observations.

The unit of absorbed dose is defined by the ICRU (see Chapter 10) as the energy absorbed (E) per unit mass (m).

$$D = E/m$$

In SI units this is Joules per kilogram (Jkg^{-1}) and is given the special name Gray (Gy). When prescribing a dose of 50Gy it could in principle also be prescribed in fundamental units as 50 $J kg^{-1}$. The Gray is small when compared with other more obvious examples of energy dissipation. For a tumour of mass 100g (0.1 kg) a total dose of 50Gy results in a deposited energy of only 50 x 0.1 = 5 Joules. This is comparable to that delivered to a 1KW heater in only 5 milli-seconds. The heating effect of ionizing radiation per Gy is thus very small with the temperature rises in irradiated tissues resulting from typical clinical doses being less than 0.001°C (0.00024°C per Gy for water). Calorimetry, the measurement of such temperature rises, although technically challenging is the only method of directly determining absorbed dose. This methodology is discussed later. We know that there is a clear relationship between tumour control probability (TCP) and absorbed dose so it is best to quantify the delivery of radiotherapy by prescribing in terms of absorbed dose. Accurate and traceable dose determination is crucial in maintaining consistency between different treatment equipment within a single department as well as maintaining consistency between equipment in different departments nationally and internationally. Clinical trials and the adoption of clinical protocols implicitly rely on the consistency of dose measurement.

7.1.2 Exposure

The way in which absorbed dose has been quantified has been governed by the radiation energies involved and the technology available to quantify them. The first radiation quantity to be defined was exposure. This concept relies on the number of ionization events measured as an indication of deposited energy in a medium. The greater the energy deposited, the greater the dose. The definition of exposure has undergone many refinements over the years, but the most recent definition is:

> The unit of exposure (X) is defined as the quotient Q/m where Q is the total charge of the ions (of one sign) produced in air when all electrons liberated by photons in air of mass m are completely stopped in air.

$$X = Q/m.$$

In SI units this is Coulombs per kilogram (Ckg^{-1}) where coulomb is the SI unit of charge.

The definition of exposure specifically mentions air as the material in which the ions are produced and completely stopped and concerns only photons. Primary standard free air ion chambers operating in the kilovoltage X-ray range try to adhere to the above definition as closely as possible. It should noted that the effective atomic number (Z) of air (7.64) is similar to that of water and soft tissue (7.42). This is important when considering measurements in low kV energy ranges when the photoelectric effect is dominant (proportional to Z^3).

7.1.3 Kerma

Photons are not directly ionizing. They do not possess charge and are considered to only transfer energy to the irradiated material via Compton, photoelectric or pair production processes. These interactions produce a charged particle, be it a recoil electron (Compton), liberated electron (photoelectric) or electron and positron pair (pair production). The uncharged photons transfer kinetic energy to these charged particles via these interactions. In the second stage the charged particles then impart energy to the material through collisional losses and a small fraction of radiation losses. This fraction is generally small compared to collision losses and will not be discussed in detail.

Kerma was introduced to describe the first part of this two stage process and is defined as:

$$K = E_{tr}/m$$

where E_{tr} is the sum of the initial kinetic energies of all the charged particles liberated by uncharged particles in a mass m. Kerma is an acronym for Kinetic Energy Released per unit Mass. The unit of kerma is Joules per kilogram (J kg^{-1}), which, as for absorbed dose, is given the special unit of Gray (Gy). Kerma differs from exposure in that it can be defined for any material not just air. A statement of kerma is not complete without defining the material concerned.

7.2 The relationships between exposure, kerma and absorbed dose

7.2.1 Exposure and kerma

The concept of exposure is easier to understand than kerma. However, exposure and air kerma are very similar with air kerma essentially being the expression of exposure in terms of energy rather than charge. Air kerma concerns the *transfer* of energy whereas exposure concerns the *absorption* of energy, required to liberate a certain amount of charge via ionization of air molecules. Exposure can be measured directly but kerma cannot. However, the relationship between exposure and air kerma is straightforward and indicates how air kerma can be derived from an exposure (charge) measurement.

$$K_{air} = X. \ (W/e) \ Gy$$

W/e is a constant of value 33.97 J C^{-1} and represents the energy required to create a charge of 1 C in air.

Determination of air kerma is all very well but few patients are made from air! From a dosimetry perspective the ideal patient is one made entirely from water and so we are more interested in water kerma.

Before looking at the relationship between kerma and absorbed dose we need to introduce two more quantities. A source of ionizing radiation gives rise to a radiation field.

Particle and energy fluence are two such quantities used to characterize the radiation field. Within the radiation field there will be a flow or fluence of particles (Φ), defined as the number of particles (N) incident on a sphere of cross sectional area (a). Fluence is defined as:

$$\Phi = N/a \qquad SI \ units \ m^{-2}$$

A sphere is chosen rather than a plane area so that the presented cross sectional area is the same from all directions. Particle fluence is therefore independent of radiation direction.

Instead of just the number of particles we could consider the energy carried by these particles. The energy fluence (Ψ) is defined as the radiation energy (R) entering a sphere of cross sectional area (a):

$$\Psi = R/a \qquad SI \ units \ J \ m^{-2}$$

7.2.2 Kerma and absorbed dose

Absorbed dose concerns the energy absorbed per unit mass to a medium whereas kerma concerns the energy transferred from photons (or neutrons) to charged particles within that medium. Considering a thin layer of material, represented by the

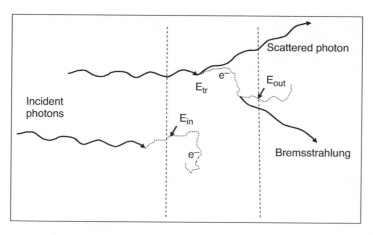

Fig. 7.1 Relationship between kerma and dose.

dotted lines in Fig. 7.1, we can demonstrate the problem that needs to be resolved to equate kerma and absorbed dose. In this explanation we will only consider photons as the incoming radiation, and electrons as the charged particles. Incoming photons may interact at any point within the medium. The kinetic energy released at the interaction point is represented by E_{tr} and we can see that, for an interaction in the layer between the dotted lines, some of the energy transferred will be expended through collisional losses and absorbed within this layer but a proportion represented by E_{out} will be imparted beyond this layer. The amount of energy E_{out} does not contribute to the absorbed dose within the dotted layer. For a single photon interaction this represents a problem as kerma and absorbed dose will never be the same in this region. However, in a realistic situation there will be many similar photon interactions occurring throughout the irradiated volume.

Some interactions occurring outside the region of interest result in energy being deposited within the layer (E_{in}). On average the kinetic energy lost (E_{out}) from electrons leaving the thin layer will be balanced by the kinetic energy gained from electrons entering the thin layer i.e. $E_{out} = E_{in}$. The absorbed dose or net kinetic energy imparted to this layer can be written as:

$$Dose = Total\ energy\ transferred - E_{out} + E_{in}$$

If we assume that the condition described above holds and that the total energy in and out are the same we arrive at:

$$Dose = Total\ energy\ transferred$$

By equating the charged particle energy entering to that energy leaving we have assumed that a situation called **charged particle equilibrium** (CPE) exists in the region of interest. So, on the assumption that CPE exists we arrive at the conclusion that the absorbed dose in a medium (D_{med}) is equal to that energy transferred, but subsequently absorbed locally through collisions (K_{coll}).

$$D_{med} = K_{coll} \qquad assuming\ CPE$$

Under conditions of CPE absorbed dose is equal to collision kerma. For X-ray energies below about 300keV the radiative losses due to bremsstrahlung are negligible (i.e. Krad ≈0) so that absorbed dose and kerma are taken as the same. However the relationship between kerma and dose becomes more complex as the photon energy increases to the MV range. The concept of CPE as expressed here is somewhat simplified and is difficult to achieve in MV beams. The closest we get to CPE is at the point of dose maximum. Beyond this, the kerma and dose curves separate by a distance that depends on the beam energy and a condition known as transient charged particle equilibrium (TCPE) exists. TCPE can be used instead of CPE with suitable correction factors. The relationship between kerma and dose is discussed further in Chapter 8 (X-ray beam physics).

7.2.3 Electronic equilibrium

The special condition of CPE (also called electronic equilibrium) is an important concept in radiation dosimetry. Electronic equilibrium exists if the number of electrons entering a space is the same as the number exiting and if their energies are similar. Why is electronic equilibrium so important and what are the consequences when it does not occur?

Bragg-Gray cavity theory allows us to perform a dose measurement in a medium with a detector and then calculate the dose to that point in the medium if the detector were not present. For example, if more electrons leave than enter a detector volume because of the design of the detector then the 'dose' calculated from the detector's signal will not represent the dose at the same point of the detector were not there. The disruption in CPE is most frequently caused by a step change in the material the photons are interacting with. This is an issue for clinical treatment plans as well as for dosimeters. A good example of this is radiation passing through muscle into bone (Fig. 7.2).

Consider the regions immediately each side of the interface between the two different materials in Fig. 7.2. They are close enough to each other that we can assume that the number of photons per unit area is the same. The kerma on each side of the interface will differ depending on how readily photons interact with each material; the more interactions that occur, the more electrons are set in motion and the kerma is higher. This depends on things such as the photon energy as well as atomic number and electron

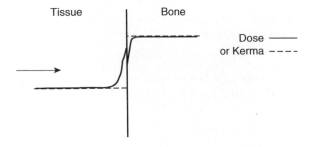

Fig. 7.2 Dose and kerma at a tissue interface.

density of the materials considered. This change in kerma is in proportion to the mass energy transfer coefficients (see Chapter 2) of the two materials and therefore produces a step change in kerma at the boundary. As the kerma changes, so does the absorbed dose on each side of the boundary, this time according to the ratio of the mass energy absorption coefficients of the two materials. Note that while the kerma curve changes in a discrete step, the dose changes more gradually. The increase in dose in muscle close to the bone is due to electrons being back scattered from the higher density bone to the right. The increase in dose in bone is due to the build up effect, similar to when a photon beam passes from air to soft tissue and described in Chapter 8. Such interfaces provide examples of electronic disequilibrium. Interfaces are clinically important in treatment planning but are the hardest to model accurately. Many computer planning algorithms do not correctly model these subtle build up/build down dose changes close to interfaces leading to considerable inaccuracies in these regions which become more pronounced at higher energies. Practical examples of this include secondary build up in the superficial layers of the larynx between the air and larynx wall. A similar effect can also occur for lung patients, particularly if a large volume of low density lung is traversed immediately prior to the tumour. This particular example is discussed further in Chapter 10. To reduce these effects a lower beam energy (e.g. 6MV) would be chosen.

7.3 Absorbed dose in different tissues and materials

7.3.1 Mass energy absorption coefficients

The energy absorbed per unit mass in two different materials, subjected to the same photon fluence, will be in proportion to their mass energy absorption coefficients. The ratio of mass energy absorption coefficients is:

$$(\mu_{en}/\rho)_{med}/(\mu_{en}/\rho)_{air}$$

If we can determine the absorbed dose in one material, such as air, we can convert this to the absorbed dose in another material such as water by multiplying by the ratio of their respective mass energy absorption coefficients. In Section 7.2.2 we showed that dose to air is equal to kerma when we have CPE. So we have now related exposure to air kerma to dose to air. Therefore, to convert dose to air to dose to water we use the equation below:

$$D_{water} = D_{air} (\mu_{en}/\rho)_{water}/(\mu_{en}/\rho)_{air}$$

For materials with similar atomic number to air the ratio $(\mu_{en}/\rho)_{med}/(\mu_{en}/\rho)_{air}$ varies gradually and by not very much with energy. For water and air this ratio is approximately 1.1 between photon energies of 100keV and 10MeV. This is very fortunate as the distribution of photon energies (spectrum) does not need to be known precisely in order to accurately determine absorbed dose from a measurement of exposure.

For materials such as bone, which has a higher atomic number than air, the ratio can vary dramatically with energy. At lower energies the photoelectric effect is more probable in higher atomic number materials (proportional to Z^3) than in air and so the ratio of relative absorption will be large. As the energy increases into the Compton region electron density becomes the important parameter in determining the interaction

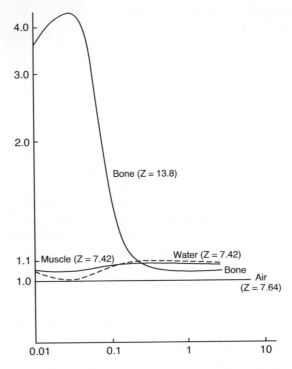

Fig. 7.3 Variation of $(\mu_e/\rho)_{med}/(\mu_{en}/\rho)_{air}$ with photon energy. Adapted from '*The Fundamental Physics of Radiology*', Meredith and Massey, 3rd Edition, J Wright (1977).

probability making the ratio approximately constant. We can also see from this that at low energies bone can absorb over four times the energy per unit mass than water. For this reason electrons may be favoured in such cases.

7.4 Methods of radiation measurement

7.4.1 Requirements of a dosimeter

Radiation produces a number of effects and quantifying the effect can make a measure of dose. In principle, any effect could be used as the basis for dose measurement as long as the relationship between the measured effect and absorbed dose can be determined.

To perform well, a detector would fulfil a number of requirements. It would:

- Be sufficiently accurate across the range of doses used in modelling and treatment in radiotherapy;
- Have a precision appropriate to resolve small dose differences for the same purpose;
- Be sensitive (high signal for small dose and not be subject to noise problems);
- Be linear across the dose range used;
- Be independent of dose rate;
- Have a response which is independent of dose (to measure large or small doses equally well);

- Have a response which is independent of energy;
- Must be able to represent dose in tissue and not just the material of the detector;
- Be small enough to have spatial resolution for use in high dose gradients.

No detector fulfils all of these requirements and different detectors are chosen for different measurement situations.

7.5 Ionization chamber devices

7.5.1 Ionization

If sufficient energy is transferred in an interaction of radiation with matter, ion pairs are formed. This ionization is caused by an electron absorbing sufficient energy that its energy then exceeds the binding energy of its atom and becomes free. It leaves behind a positively charged atom resulting in an ion pair. The amount of ionization produced is proportional to the energy delivered to the medium and so measurement of the amount of ionization can be the basis of not only a radiation detector but also of dose measurement. Ionization chambers usually measure charge created in air.

7.5.2 Principles of operation

7.5.2.1 The need for applied voltage

In order to measure the number of ions produced they must be collected. This is done simply by attracting them to an electrode of opposite charge and counting them in a device called an electrometer. In travelling through the medium, the ions produced will be attracted to any ions of opposite charge in the vicinity and there will be a tendency to recombine. If recombination occurs the ionization event cannot be detected and hence will not contribute to the measurement and the dose will be underestimated. To minimize the chance of recombination, the potential difference (voltage) between the electrodes must be sufficient to collect the electrons quickly. The collection of electrons constitutes a current through the electrometer which can be measured by a current meter. As the voltage is increased, the number of electrons collected (i.e. prevented from recombining) increases until all of the available electrons are being collected. Care must be taken to minimize the leakage of charge collected from the chamber assembly by use of appropriate insulation materials.

7.5.3 Types of ionization chambers

7.5.3.1 Free air ionization chamber

The free air chamber is a primary standard designed to determine exposure for photon energies up to around 300kV. Free air chambers are maintained by national standards laboratories and used to calibrate reference standards, which in turn are used to calibrate the routine equipment used in the clinic (see later). The free air chamber is designed to comply as closely as possible with the definition of exposure and the conditions to satisfy charge particle equilibrium and is shown schematically in Fig. 7.4 and the UK National Physical Laboratory's free air chamber is shown in Fig. 7.5. It consists of an air filled metal box with an aperture allowing a well defined area (A) of X-rays to pass through the chamber so that it only interacts with air within the chamber. Ions created

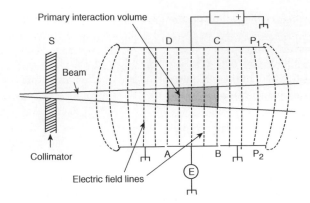

Fig. 7.4 Schematic diagram of a free air ionization chamber. The guard plates outside the AB plate exclude any signal from the region where the electric field lines are bowed to ensure that the volume from which ionization is collected is known.

when the X-rays interact with the air are attracted toward two high voltage electrodes placed on opposite sides of the box. The electrode separation is sufficient to ensure that electrons emanating in the shaded region lose all their energy before they reach the electrodes i.e. they are completely stopped in air. Electronic equilibrium is established as long as the shaded volume is sufficiently far inside the box that full electron build up has been achieved. This means that electrons produced in but travelling out of the dotted region and not collected by the electrodes are compensated for by electrons entering this region from outside and subsequently collected by the electrodes. The total electron charge (Q) is measured and the mass of air calculated

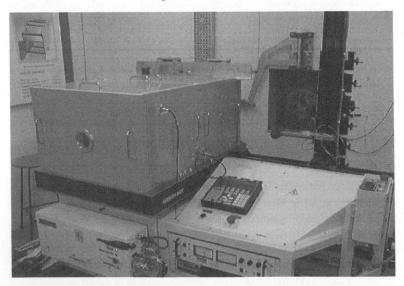

Fig. 7.5 The free air ionization chamber at the UK's National Physical Laboratory. Note the round beam entry window in the steel casing. Reproduced with permission from the National Physics Laboratory.

from the air density (ρ) and dimensions of the shaded volume. Several small correction factors are applied to account for such things as X-ray attenuation in air between the entrance aperture and shaded collection region, the presence of any water vapour, scattered radiation entering the chamber from outside, ionization from bremsstrahlung, not all electrons being collected by the positive electrode etc. Although basic construction is similar the medium energy free air chamber is substantially larger ($1m^3$) in order to satisfy the definition of exposure and to ensure electronic equilibrium is reached at the measurement volume.

As the beam energy increases, the separation of the plates must increase to prevent electrons reaching the plate before giving up all of their energy.

7.5.3.2 Thimble chambers

More useful in the clinic is a relatively small ionization chamber, usually thimble shaped, which encompasses a typically $0.1\ cm^3$–$1.0\ cm^3$ cavity within which ionization occurs (the active volume) and is collected between the chamber's axial electrode and its conducting walls. Such a chamber is shown in Fig. 7.6. The walls are close to tissue or air equivalence using a low-Z material such as graphite or conducting plastic. The central electrode is usually aluminium. This has a relatively high atomic number but the effective atomic number of the detector as a whole is similar to air. The commonly used Farmer type thimble chamber has an external diameter of 7mm and a 1mm Aluminium electrode and a 0.5mm graphite external wall making the gap between the electrodes 2.5mm. In practice 200V is used to guarantee collection of virtually all ions created. The thimble is one of the key components in a dosimetry system as it accurately defines the cavity volume in which ionization of the contained air mass occurs as well as forming one of the electrodes. The material the thimble is made from is important in determining the properties of the chamber. The standard Farmer type cavity chamber is useable across the full range of radiotherapy energies from around 50kV to 25MV.

The theoretical basis on which the cavity type chamber can be used in the kV energy range (to 300kV) is quite distinct to that in the MV energy range. At kV energies the thimble material is assumed to be air equivalent, whereas in the MV energy range it is assumed to be water equivalent. A detailed description is beyond the scope of this chapter, but a brief overview is given as follows for completeness.

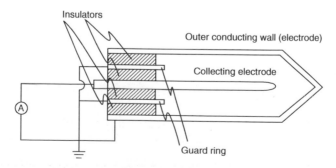

Fig. 7.6 Example of a simple thimble ionization chamber. The conducting wall is shown and axial central aluminium electrode. The wall and electrode are separated by an insulator.

For the free air chamber described above, if it were possible to compress the air around the shaded measurement volume so that electronic equilibrium still holds by say a factor of 1000 without affecting the shaded volume, the electron ranges in this compressed volume would also be reduced by a factor of 1000. If the electron fluence, i.e. the number, direction and energy of electrons leaving this compressed volume were identical to that in air then the conditions of charged particle equilibrium would be satisfied. Suitable thimble material (graphite, conducting nylon) having similar properties to 'very dense' air would then enable exposure measurement by simulating a free air chamber.

7.5.3.3 Parallel plate chambers

One disadvantage of the thimble chamber is its inability to pinpoint the position of a dose measurement in a high dose gradient radiation field. Plane parallel chambers allow the measuring point to be much better defined in space and to have a finer measurement resolution in one dimension. A parallel plate chamber consists of two conducting plate walls only one of which usually is the beam entrance wall as shown in Fig. 7.7. Parallel plate chambers are recommended for measurements in electron beams and surface and build up dose measurements in photon beams.

7.5.4 Electrometer

7.5.4.1 Principles of operation

In order to convert ionization produced in a chamber into an indication of exposure, the ionization must be collected and measured. To quantify the rate of exposure, the rate of flow of charge (i.e. current) through the circuit should be measured with an ammeter. To quantify the total exposure, all of the charge produced in the cavity (which, when moving, constitutes the current) needs to be collected and measured. This is done by storing the charge in a capacitor until the end of the exposure and then reading the potential difference across it. The current produced by a 0.6cm^3 ionization chamber is typically 10^{-9}A or, for small volume pinpoint ionisation chambers, as low as 10^{-12}A. Electrometers may also be used to assess dose rate, rather than dose.

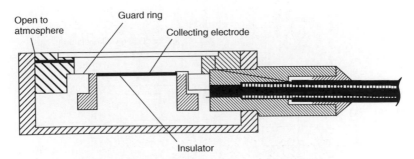

Fig. 7.7 Plane parallel chambers. The measuring point is typically immediately inside the upper front face electrode. The guard ring defines the collecting volume.

7.6 **Other measurement devices**

7.6.1 **Solid state detectors**

Detectors based on solid materials have a great advantage over gas filled detectors. Their density is significantly greater than air and therefore their ability to cause interactions with incident radiation is much greater. The section below summarizes the most commonly used solid state detectors.

7.6.1.1 **Thermoluminescent detectors**

If the light output from thermoluminescent (TL) materials can be measured and amount of light produced is calibrated against absorbed dose then the TL material can be used as a dosimeter (TLD). Commonly used TL materials are Lithium Fluoride (LiF), which has an effective atomic number (8.2) close to that of tissue (7.4), and Lithium Borate ($Li_2B_4O_7$).

In single atoms, electrons can exist only in discrete energy states but the closely packet atomic structure of crystals means that interactions between adjacent atoms results in bands of allowed energies: a lower range of allowed energies termed the valence band and an upper range termed the conduction band. The energy range between, termed the forbidden region, is not available to electrons unless some impurity defines an energy state that electrons can hold. These intermediate energy states are called electron traps. In the case of LiF crystals, e.g. these traps can be created by the introduction of impurity atoms into the crystal structure.

When an electron in an atom absorbs energy from irradiation it may have its energy state raised from the **valence** band to the **conduction** band. The electron may recombine with a positive 'hole' and return to it's default 'ground' state emitting a photon of light in the process, resulting in the phenomenon called fluorescence. If an electron settles in one of the electron traps it is in a metastable state where energy is required to lift it out of its trap in order to return to its ground state, emitting a photon in doing so. In thermoluminescence, heat is be used to provide the required energy. This process is shown schematically in Fig. 7.8.

As the temperature of the applied heat is increased the likelihood of releasing the trapped electrons increases and so the amount of light given off increases. If the crystal

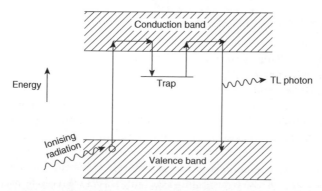

Fig. 7.8 Energy level model of thermoluminescence process.

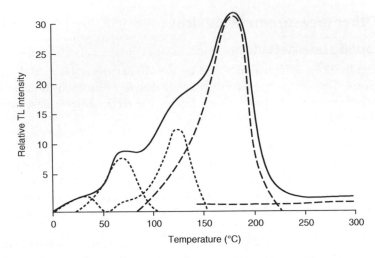

Fig. 7.9 Glow curves and dose curve used for thermoluminescent dosimetry.

contains traps of different energies then different amount of heat are required to induce the emission of TL photon from each. The amount of light given off as a function of temperature when heat is applied shows a non-linear variation and is called a **glow curve** which may contain a number of peaks associated with the different energy traps. The total light emitted is equivalent to the area under the glow curve and this can be calibrated with dose to produce a TL dosimeter. For a large range of radiotherapy energies the light output is a linear function of dose. Before use, any existing (or residual if previously exposed) trapped electrons (and holes) must be removed. This can be done using a heating and cooling cycle using specific temperatures and rates, a process called annealing.

TL dosimeters are useful for in-vivo measurements because they are small, typically 3mm square by 1mm thick, and do not require connection to measuring equipment with wires. Their main drawback is that their accuracy is limited to typically 5% or if exceptional care is taken, as good as 3%. The dose response of TLDs is dependent on their anneal history and so batches of TLD chips should be have the same anneal history.

7.6.1.2 Silicon diode detectors

A diode dosimeter consists of a silicon (Si) crystal with two regions; a p-region with an excess of positive charge (or deficit of electrons leaving positively charged 'holes') and an n-region with an excess of electrons. In practice this is produced by taking a p-type Si crystal and introducing atoms of another material in a process called doping. At the junction of these regions, some local diffusion of electrons causes some of the holes in the p-region adjacent to the junction to be filled with electrons from the n-region. This results in a third region which is depleted of mobile charges and termed the depletion layer. The p- and n-regions then cause a small intrinsic potential difference (~0.7V) across this depletion layer and any free electrons entering it will be swept across to the p-region, thus constituting a current. This current can be measured in an electrometer without the need for an applied bias voltage. The active volume of the diode detector is therefore very small in this direction ($\approx 100\mu m$). The diameter of the diode itself is of the order of 1–2mm but dosimeters are encapsulated in waterproof material and, depending on their use, added build up material making their outside dimensions typically from 3mm to 10mm.

The high sensitivity of diodes means that the detecting volume can be reduced allowing them to be small, having a high spatial resolution in all dimensions. Additionally the energy to create an ion pair in silicon is about 1/10th that in air so, for the same energy released, the generated current is an order of magnitude greater. The signal from a very small volume diode detector is therefore as great as a much larger air ionization chamber. This makes them particularly useful in measuring beam profiles because they are able to accurately measure the rapidly changing dose in the penumbra region.

The sensitivity and resolution advantages are not enough to make a diode the ideal detector. One disadvantage is a change in sensitivity with repeated use due to radiation damage. This means that they should not be used for reference or absolute dosimetry. Even for relative dosimetry they need to be calibrated at a frequency consistent with the observed rate of sensitivity change. Modern diodes are often pre-irradiated to a large dose (kGy) which, although reducing their initial sensitivity due to radiation damage, makes their subsequent change in sensitivity with accumulated dose less pronounced.

Diodes are often used to check patient entrance doses and are produced with build up material which is energy specific. Diodes exhibit a response variation with temperature, beam energy spectrum, dose rate and angle of incidence and so careful use is required to get good performance. Temperature variation may be particularly difficult to deal with if the temperature of the diode is lower than the patient's skin when placed and increases in temperature during the exposure. Ideally the diode should be at the skin temperature at the point of measurement as well as calibration. Recent commercial diodes exhibit less sensitivity variation with temperature (typically only 1–2% increase in response between room and skin temperature (around 32°C)).

7.6.1.3 MOSFET detectors

Metal oxide semiconductor field effect transistor (MOSFET) detectors are very small detectors with high spatial resolution. Because of their size they cause very little attenuation of the beam. Ionizing radiation causes electrons to be permanently trapped in the oxide layer which results in a change in the threshold voltage of the transistor. Because of the permanence of the trapped electrons, MOSFETs have a limited life as a dosimeter. MOSFETs do not require a dose rate correction but they do have a small variation in response with energy. Like diodes, MOSFETs have a temperature dependence.

7.6.2 Chemical detectors

Radiation can cause chemical changes in some materials which, if calibrated, can be used in radiation measurement systems. The most common of these is film darkening or colour change.

7.6.2.1 Radiographic film

Radiographic film consists of a thin flexible plastic sheet coated with a radiation sensitive emulsion containing silver bromide. When exposed to radiation, ionization causes a latent image on the film that becomes visible, and can be permanently fixed, by chemical processing. The greater the exposure to ionizing radiation the darker the resultant film and the less light transmission it allows. Thus the spatially variant opacity allows the original exposure variation to be viewed by shining a uniformly illuminated light box through the film. For dosimetry, light transmission is measured, using an optical densitometer, in terms of optical density which is a logarithmic function of the light intensity measured with (I) and without (I_0) the film at the position of measurement. Optical density (OD) is defined as $\log_{10}(I_0/I)$. For radiographic film to produce accurate dosimetry a carefully set up and well maintained darkroom facility for development and processing have to be maintained. Quality control of the development and processing systems must be performed and variations between individual films and batches have to be accounted for. The high Z of the film active layer causes energy variation with respect to tissue which requires correction. The advantages of radiographic film are:

1. It provides a permanent record;
2. High spatial resolution.

The disadvantages are:

1. Non-linearity of response;
2. Energy dependence at low energies.

7.6.2.2 Radiochromic film

Radiochromic film, as the name suggests, changes colour upon exposure to radiation. It is a self-developing film and the colour change is caused by polymerization of dyes embedded in an emulsion layer coated on a substrate. Once produced, the polymer absorbs light and so, like radiographic film, the transmission can be measured using a densitometer. The absorption peak depends on the film type used and the best contrast will be achieved if wavelengths around this region can be extracted. The useful dose range of the currently available film is in the region 0.01Gy to 8Gy. Although the film does not require chemical development, the post exposure dose image develops over time and it should be left for typically 6 hours before readout.

Radiochromic film has many advantages over radiographic film:

♦ No dark room required;
♦ Response independent of dose rate;
♦ Nearly tissue equivalent.

To achieve precision in the range of 3%, radiochromic film still requires a carefully set up and well maintained scanning and processing system and dose calibration process.

7.7 Detectors for dose measurement

7.7.1 Absolute dosimetry

Absolute dosimetry refers to the direct measurement of dose and is seldom performed outside of standards laboratories. For kV energies, a free air chamber is used but for megavoltage energies it is most often based on measurement of temperature rise in a graphite calorimeter. The temperature rise is tiny (pico degrees) and is only measurable with specialist equipment.

7.7.2 Reference dosimetry

Reference dosimetry is the determination of absorbed dose using a well defined standard setup and using approved reference standard electrometer and ionization chamber which is cross calibrated against a primary standard. In the UK, the approved secondary standards, known as designated transfer instruments, are all ionization chambers which demonstrate a high standard of stability with respect to variations in energy, dose and dose rate. The chambers used for external beam photons are typically Farmer type, with parallel plate type chambers for electrons and well type for brachytherapy sources.

7.7.3 Relative dosimetry

There are many dose measuring devices which do not fulfil the stringent requirements of reference dosimetry. These can still be very useful in situations where we are interested in how dose changes rather than actual dose measurements. These measurements are useful if the same corrections would be appropriate for both measurements. In this case it is unnecessary to actually make the ionization to dose corrections and indeed the required corrections may not be known. Examples of such measurements are depth dose measurements such as PDD or, field size factors etc. In choosing a device for relative dosimetry it must be known what quantities are changing between the relative measurements and a device which is insensitive to this change can be selected.

7.8 Relationship between ionization and dose

7.8.1 Phantom materials

The reference material for performing dosimetry measurements is either water or a water equivalent material. A water equivalent material is designed to have the same radiation interaction characteristics as real water over the energy spectrum at which the measurements will be made. Commercial solid water equivalent plastic materials consist of a homogenous mixture of several materials combined in appropriate proportions to produce an accurate match with water. These may also include materials to modify the density to that of water so that depths in the water equivalent material match those in real water. The advantage of water equivalent materials is that, although a lot more expensive than real water, they are rugged and enable quick and accurate set up.

7.8.2 Dose measurement in the clinic

At MV energies electron ranges are much greater and even the 'very dense' air thimble would have to become thick to provide electronic equilibrium leading to unwanted attenuation of the photon beam. At these energies a theoretical approach called Bragg-Gray cavity theory is adopted. First we need to imagine a small air filled cavity introduced into a material uniformly irradiated by photons in which charged particle equilibrium exists. If the introduction of this cavity does not modify the electrons in anyway then the electron fluence passing through the air in the cavity would be the same as in the material in the absence of the cavity. If the same number and energy of electrons pass through the material and air cavity then the ratio of electron energy lost per unit mass equals the ratio of the mass stopping powers of the material and gas concerned. So by measuring exposure and deriving the dose to air we can then calculate the dose to the surrounding material by multiplying by the ratio of the mass stopping powers for the material, usually water, and air.

7.9 Calibration and traceability of dose determination

7.9.1 The calibration chain

The requirement of all dosemeters is that they exhibit some response (R) to radiation. To use that response as a measure of dose requires an additional factor, the calibration factor (F) that under the conditions in which the device was irradiated will convert the response to absorbed dose (D).

$$D = F \times R$$

For equipment used in the clinic this factor is routinely determined by comparing the uncalibrated device with another device that is already calibrated. The calibration factor is then that multiplier required to make the uncalibrated reading match the known dose as shown by the calibrated device.

This process of transferring the calibration factor is called an intercomparison or cross calibration. This process must be performed with care. Dosimetry Codes of Practice (CoP) describe in detail how this is carried out. CoPs provide definitive guidance on:

i) Performing a cross calibration between a calibrated secondary standard and an uncalibrated dosemeter;

ii) Conducting an absorbed dose measurement using a calibrated dosemeter.

At the top of this calibration 'chain' sits a device that cannot be calibrated by comparing with another. The calibration chain starts with a primary standard, a device that

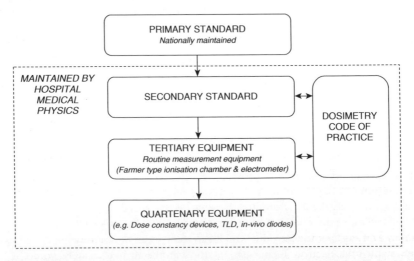

Fig. 7.10 Calibration chain from the centrally maintained primary standard down into the hospital clinic.

determines the quantity of interest from first principles. In this hierarchy all devices depend on the accuracy of the primary standard. Figure 7.10 shows this calibration chain from the centrally maintained primary standard down into the hospital clinic. Each vertical arrow indicates that a cross calibration must take place in order to disseminate the primary standard calibration down the chain. The CoP details how absorbed dose is to be measured in the clinic and also how the cross calibration between secondary standard and routine equipment is performed.

7.9.1 Dose standards

7.9.1.1 National standards

By their nature primary standards are complex pieces of equipment requiring dedicated full time staff for maintenance and operation. These standards are not transportable and are inappropriate for use in a clinical setting. They are maintained at a national level in dedicated primary standard dosimetry laboratories (PSDLs). In the UK this is the National Physical Laboratory (NPL) at Teddington, Middlesex. There are currently five primary standards covering external beam photon and electron radiotherapy maintained at NPL and these are summarized in Table 7.1.

To satisfy the conditions of charged particle equilibrium the size of a free air chamber should be around twice the range of the maximum energy electrons generated from the photons being measured. At higher energies this becomes impractical as it would require a chamber several metres across. As a result standards laboratories use air filled graphite walled cavity ionization chambers to realize air kerma at higher energies. Graphite is reasonably air equivalent but has a much higher density and therefore can be thought of as a large volume of air compressed into a much thinner rind of graphite. It relies on the fact that the volume of the air cavity within the graphite is accurately determined. For a known volume and atmospheric conditions the precise mass of air inside the small cavity can be calculated. The charge liberated by this known mass of air can then be used to calculate the dose to air. Air kerma can then be derived from this dose to air using additional factors and Bragg-Gray cavity theory.

Charge liberated in a medium by ionizing particles result in an energy cascade in which energy is shared amongst many secondary particles. Eventually these less energetic charges and ions recombine with the resultant energy being liberated as heat. Although the number of ionizations generated per treatment fraction is sufficient to

Table 7.1 Five primary standards covering external beam photon and electron radiotherapy

Primary standard	Energy range	Quantity realized
Free in air	8–50kVp	Air kerma
Free in air	40–400kVp	Air kerma
Cavity air ionization chamber	2MV	Air kerma
Graphite calorimeter (photons)	4–19MV	Absorbed dose to water
Graphite calorimeter (electrons)	4–25 MeV	Absorbed dose to water

cause cell death the energy involved is very small. Calorimetry is the technique of deriving absorbed dose via a measurement of temperature rise. The principle of calorimetry is straightforward but the practical realization is difficult due to the small temperature rises involved. The absorbed dose in an irradiated medium is given by the measured temperature rise multiplied by the specific heat capacity of the medium. The specific heat capacity is the energy required per unit mass of material to increase its temperature by 1°C.

$$D_{med} = C_{med} \times \Delta T$$

where:

$$D_{med} = \text{Absorbed dose to the medium (J kg}^{-1}\text{)}$$

$$C_{med} = \text{Specific heat capacity of the medium (J kg}^{-1} \, °C^{-1}\text{)}$$

$$\Delta T = \text{Temperature rise (°C)}$$

The assumption in the equation above is that no other physical or chemical changes occur in the irradiated material that may result in absorbed energy not being manifested as a rise in temperature. Graphite is currently used for primary standard calorimeters. Compared to water it is a rugged stable material with a temperature rise six times that of water for the same dose (0.0014°C/Gy) for graphite. The downside of using graphite is that the measured dose is to graphite and must be converted to that in water.

Compatibility of measurements in different countries depends on the consistency of their respective national standards. This is tested by comparing primary standards directly or via an intermediary such as the International Bureau of Weights and Measures (BIPM) in Paris who coordinate the international measurement system. Internationally there are currently four different methods in use for establishing absorbed dose to water and these display a remarkable agreement (<1%). The consistency of air kerma standards is not as good.

7.9.1.2 Local standards

The role of the PSDLs is to provide a calibration factor for the secondary standard. This local standard is maintained by one or perhaps a group of several hospital physics departments and sent to the primary standards laboratory typically every three years for recalibration. The local standard is normally only used to cross calibrate other 'tertiary' equipment, typically annually, used for routine measurements. The main use of routine equipment is to calibrate the output from the treatment equipment and then to confirm machine stability over time by repeat measurements at fixed intervals. The term output means absorbed dose (Gy) per set machine monitor unit (or time for some equipment) under appropriate reference conditions. Additionally tertiary equipment will be used to calibrate other quaternary equipment such as dose constancy check devices, in-vivo diodes, TLD etc. Secondary standard equipment is built to a higher quality and performance specification than equipment used for routine measurements and requires careful maintainance so its calibration factors remain valid over the periods between subsequent recalibration at the PSDL.

7.9.2 **Dose measurement corrections**

The equation $D = FxR$ described above to determine the dose (D) based on a dosemeter response (R) and calibration factor (F) requires the dosemeter to be subsequently used under identical conditions at which it was calibrated for the calibration factor to be valid. Possible variations in conditions are many and some are listed below:

1. Temperature,
2. Pressure,
3. Humidity,
4. Detector orientation,
5. Depth of measurement,
6. Total dose delivered,
7. Dose rate (dose per linac pulse),

The less sensitive a detector is to any of the above the better, however all detectors are sensitive to one or more. This means that the response of the detector must be corrected to the conditions under which it was calibrated.

Chapter 8

X-ray beam physics

R Mackay and A Hounsell

8.1 **X-ray beams used in clinical practice**

In external beam radiotherapy, the radiation originates in a machine some distance from the patient surface. The properties of the X-ray beam depend on what and how the radiation is produced. X-rays are only produced when the 'beam is on' and are the result of the collision of accelerated electrons with a target material and thus X-rays are bremsstrahlung radiation. An important determinant of beam energy is the electrical potential through which a beam is accelerated. When an electron is accelerated across a voltage of 50,000V it will acquire energy of 50,000eV and this is the maximum energy that can be transferred to the X-ray produced by the bremsstrahlung interaction of the electron with a target. Often the energy on a treatment machine will be referred to by a kilovoltage (kV) or megavoltage (MV) potential that represents this maximum possible energy: in reality most of the photons will have less energy than this maximum and the spectrum of energies of an X-ray source will have a peak at approximately one third of the maximum (see Chapter 3, Fig. 3.1).

The shape of the photon energy spectrum of the X-ray beam depends on the target material, filtration of the beam and design of the X-ray head. However, the penetration of the beam is related to the maximum accelerating potential. X-ray treatment machines range from superficial X-ray units, designed only to treat to a depth of <5mm, to linear accelerators intending to treat tumours in the middle of the body.

Although megavoltage strictly applies to any beam over 1MV in practice, radiotherapy beams typically range from 4MV to 25MV with most departments having a 4MV–8MV option and a higher 10–18MV facility.

8.2 **Beam quality and quality indices**

The energy of the beam or beam quality is a measure its ability to penetrate a material—the higher the energy the more penetrating the beam is. However beams of the same energy may penetrate a material to different amounts. This is due to small differences in the energy of the accelerated electrons, the thickness and composition of the target, effect of any added filtration and the design of the beam defining system.

It is important to specify the energy of a beam for several reasons:

i) To be able to predict the penetrative characteristics of the beam;

ii) For use with dosimetry protocols in which data from standard laboratories are specified in terms of energy (see Chapter 7);

iii) To allow the comparison of treatments units and the outcomes of clinical studies between different centres (see Chapter 16).

There are a number of ways of describing the beam quality and these are discussed next.

Table 8.1 Range of use of different radiotherapy treatment units

	Accelerating potential	Clinical treatment depth	Clinical use
Superficial	50kV–160kV	<5mm	Skin lesions
Orthovoltage	160kV–300kV	<6cm	Shallow targets e.g. skin, superficial tissues, and ribs
Megavoltage	>1MV	<30cm	Deep seated tumours e.g. prostate

8.2.1 Quality indices for kilovoltage X-rays

The particular quality index used to describe a beam is in turn dependant on the energy of the beam. At kV energies the half value layer (HVL) (see Chapter 3) is used i.e. the thickness of a specified absorber which reduces the beam intensity to half its original value. It describes the ability of the beam to penetrate a material and is therefore clinically relevant. The HVL is obtained by measurement of an absorption curve using the specified material. Sometimes beams of different spectral distributions can have the same first HVL but different second HVL values. This is due to the beam being comprised of a spectrum of X-ray energies which are differentially attenuated. The ratio of the first to second HVL is termed the homogeneity index and is 1 for a monoenergetic beam and less than 1 for heterogeneous beams. When measuring HVL the use of good geometry and the reduction of scattered radiation is important. When a broad beam is incident on sheets of attenuating material there is additional scattered radiation in addition to the primary radiation. This results in a seemingly more penetrating beam, the magnitude of that increase depending on the field size (Fig. 8.1).

Kilovoltage and orthovoltage beams are specified in terms of their HVL, their peak kV value (kV_p) and often the filtration added to harden the beam before the HVL measurement is made. Example values are shown in Table 8.2. Superficial energy HVLs are usually specified in terms of mm of aluminium, higher energies being specified in terms of mm of copper.

8.2.2 Quality of megavoltage beams

HVL in pure metals (as used for kV beams) is not suitable at MV because it is both a slowly varying function of energy and it can be affected by pair production at higher energies. Water is a more suitable material because it is tissue equivalent. There are several ways of using water to specify the energy of the beam.

8.2.2.1 Tissue phantom ratio (TPR_{10}^{20}) or quality index (QI)

This specifies X-ray attenuation by measurement of the ratio of two points on a depth ionization curve well beyond d_{max}. This avoids any problems associated with

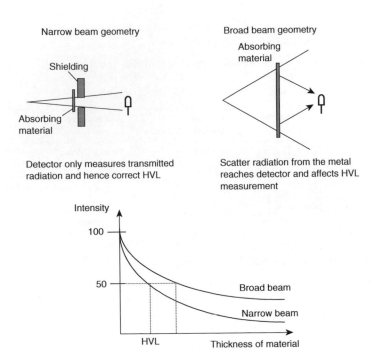

Fig. 8.1 Correct narrow, and incorrect broad, beam geometries for measuring HVL and the resulting shape of the attenuation curves.

measurements being made at d_{max} which can be influenced by electron contamination of the beam. The tissue phantom ratio (TPR) measurements use the ratio of the dose at two depths at a fixed source to detector distance. To characterize the energy, the ratio is used of the dose at 20cm deep to 10cm deep for a field size of 10cm by 10cm in water at a fixed chamber source distance of 100cm (Fig. 8.2). The difference in dose is predominantly due to attenuation of the radiation in the water because both measurements are made at the same distance which removes inverse square effects.

Although not strongly dependent on beam energy, this ratio is widely used to specify beam quality, especially for dosimetry purposes. For example, in the UK MV dosimetry chain (see Chapter 7) a hospital obtains calibration factors from the National Physical Laboratory (NPL) for its secondary standard dose meter by relating its local beam energies to beam energies measured at the NPL using the QI.

There are several other ways for specifying the beam energy.

Table 8.2 Nominal HVL values for typical kV values and added filtration values

kV$_p$	Typical added filtration values (mm Al)	Nominal HVL (mm Al)
100	1.15	2.1
70	0.75	1.1

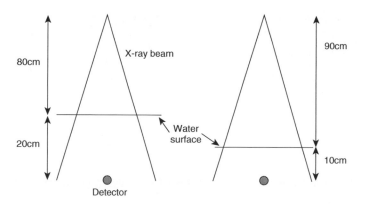

Fig. 8.2 Measurement geometries for determining the QI at megavoltage energies.

8.2.2.2 HVL and narrow beam attenuation coefficients (μ_0) in water

The HVL in water can be obtained from measured data provided that the amount of scattered radiation from the water is minimized either during the measurements or by extrapolation to zero field size. These values are difficult to obtain routinely but are used in some dose models and algorithms.

8.2.2.3 Using percentage depth dose (%dd) data

The ratio of two central axis percentage depth dose values, measured at specific depths and for a specific field size can be used as a measure of the beam energy. The surface of the phantom is at a fixed SSD, and field size is specified at the surface. This is commonly used for quality control measurements and is simple to perform but is not a very sensitive measure of energy. Typical depths used are 5cm with 15cm or 10cm with 20cm.

Another method is to use the % depth dose value at some distance beyond d_{max}. There are two ways of doing this:

i) the depth of a particular %dd value e.g. d_{80}(cm);

ii) the dose at a particular depth e.g. D_{10}(%).

The advantages of these specifiers are that they are relatively easy to measure and vary significantly over the whole quality range and are usually at clinically relevant depths (Fig. 8.3). Both these are used with linear accelerator (linac) manufactures often specifying the beam quality using the D_{10}(%) value.

BJR Supplement 25 provides detailed information about typical values for these metrics. Some example values are given in Table 8.3.

8.3 Depth dose characteristics

The dose deposited in a patient can be considered to be in three main parts:

◆ The surface dose,

◆ The build up region,

◆ The region beyond the depth of maximum dose (Fig. 8.4).

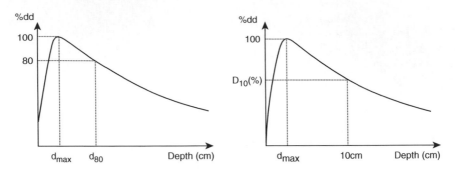

Fig. 8.3 Schematic illustrating different methods for specifying energy of a beam using the %dd.

As a photon beam enters a medium it begins to lose energy through interactions with the atoms in the material. The dominant interactions for radiotherapy are the photoelectric effect, Compton scattering or pair production. The most common interaction at MV energies is Compton scattering, which produces predominantly forward scattered electrons that travel through the medium, ionizing atoms and depositing dose. The kinetic energy released from the photon beam to the electrons is known as kerma (kinetic energy release per unit mass—see Chapter 7). Kerma is highest at the surface, where the photon beam has the greatest intensity, but paradoxically the dose is low. Kerma decreases with depth as the beam is attenuated i.e. the number of photons available to transfer energy decreases.

There is some dose at the surface from radiation backscattered from the phantom, and contamination radiation (photons and predominantly electrons) from the treatment head and air gap. The surface dose value decreases with increasing energy. This reduces the dose to skin at the entrance of a radiotherapy beam, an effect called 'skin sparing'. Skin dose increases with increasing field size and if accessory trays or physical wedges are introduced into the beam. Practically, dose at the surface is difficult to measure because of the rapid gradient of the depth dose curve, the lack of electronic equilibrium and the complexity of using standard ionization chambers which are designed to measure dose at depth.

Moving into the phantom fewer photons are available to interact and the kerma falls but the dose is demonstrating 'build up' as the electrons set in motion at or close to the

Table 8.3 Example nominal values for different methods for specifying beam energy at MV energies

Nominal MV	D_{10}(%)	d_{80}(cm)	d_{max}	QI	%dd(5)/%dd(15)
4	63.0	5.9	1.0	0.626	1.80
6	67.5	6.7	1.5	0.677	1.68
8	71.0	7.5	2.0	0.713	1.61
15	77.3	9.2	3.0	0.757	1.52

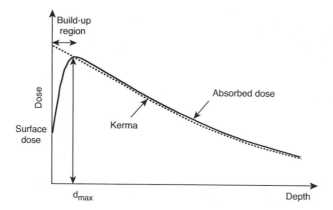

Fig. 8.4 Dose (solid line) and kerma (dashed line) as a function of depth. The build up region relates to the part of the curve before the dose reaches the maximum. At greater depths the dose falls off due to the attenuation of the beam and also the spreading of the radiation from the source.

surface come to the end of their path and deposit more energy. At a depth below the surface, d_{max}, the dose reaches a maximum. At this point, in theory, the kerma curve intersects the dose curve and we have a point of charged particle equilibrium (CPE, see Chapter 7) where the energy deposited is the same as the energy transferred. In practice, the assumption generally is invalid due to the presence of contaminating electrons generated by the beam outside the phantom between the radiation source and the phantom surface. d_{max} increases with energy, it can also vary with field size and with the addition of accessory trays or physical wedges. This is due to additional contamination radiation contributing to the dose in the build up region.

Beyond d_{max}, kerma and dose both decrease due to attenuation of the photon beam and the inverse square law. The rate at which the beam is attenuated depends on its energy—the higher the beam energy the less attenuation. However, the reduction in dose due to the beam spreading out is governed by the inverse square law which is **independent** of energy and is approximately 2% per cm at a distance of 100cm. As kerma falls and less energy is transferred so less energy is deposited. Both curves decrease in parallel, the kerma curve being lower, separated by a distance dependent of the photon energy. The higher the energy, the greater the distance travelled by electrons set in motion before they come to the end of their path and deposit most dose. The relationship between kerma and dose beyond d_{max} is constant and transient charged particle equilibrium (TCPE) is said to exist.

8.3.1 Variation with field size

The %dd characteristics of a beam change with changes in field size as illustrated in Fig. 8.5a. This is because the dose at depth in a material has a component due to scattered radiation. The amount of scattered radiation increases with increasing volume of material irradiated and this depends on surface field size and depth within the material. As the field size increases, the beam becomes more penetrating due to the increased

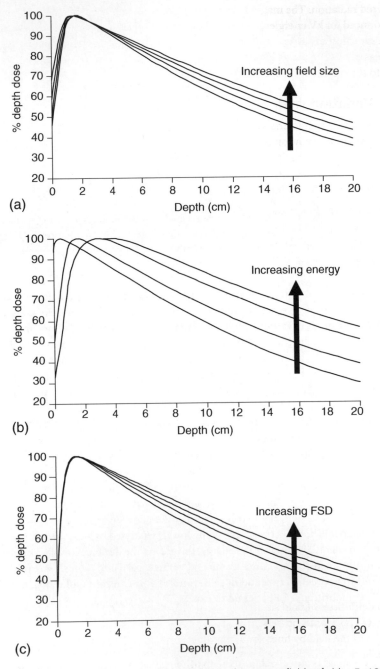

Fig. 8.5 Variation of % depth dose with a) field size for square fields of sides 5, 10, 20, 40cm at 6MV; b) with energy for Co-60, 6MV, 15MV and 25MV; c) with FSD for 80cm, 100cm 120cm and 150cm for a 10cm by 10cm field at 6MV.

scattered radiation. The magnitude of this effect is also energy dependent, being more pronounced for kV energies and less pronounced for MV energies. The reason for this is the direction of scatter produced. At lower energies, scatter occurs in all directions whereas at MV energies it is in the direction of the beam, so contributing relatively less dose to the central axis dose.

8.3.2 Variation with energy

The depth dose characteristics of a beam change with the beam energy. This is due to changes in the attenuation of the beam which is a function of beam energy. The higher the energy of the beam the more penetrating the beam as illustrated in Fig. 8.5b.

8.3.3 Variation with focus to surface distance (FSD)

The %dd characteristic of a beam changes with FSD (Fig. 8.5c). This is because the %dd contains both a component due to attenuation of the beam and an inverse square component. As the FSD increases the beam becomes more penetrating as the influence of the inverse square component reduces. Similarly as the FSD decreases the beam becomes less penetrating as the inverse square component becomes more dominant.

8.4 Methods to describe the treatment beam

In this section definitions of beam geometry and field size are considered before the concepts around isodose lines are discussed.

8.4.1 Beam geometry

It is important to have a standard definition of beam geometry. As the beam is diverging, the treatment field will increase with increasing distance from the X-ray source.

8.4.1.1 Field size definition

This is a description of a square or rectangular field using two dimensions. The field size is usually defined at the isocentric plane which is typically at 100cm FSD. The field size at the isocentric plane is also referred to as the jaw settings. Alternately the **surface** field size may be used which will differ from the jaw settings unless, of course, the surface is at the same distance as the isocentric plane. The field will diverge with increasing distance from the X-ray source and so treatments at extended or reduced FSDs will have a different surface field size than the accelerator jaw positions indicate.

Two definitions of field size can be used: the geometric field size and the dosimetric field size. The geometric field size is a projection of the front edges of the collimator system into the field by lines drawn from a point at the centre of the front face of the source. Such lines define the geometric edge of the beam. The dosimetric field size is the area enclosed by a specified isodose line. The geometric field size is equivalent to that defined by the 50% isodose while the therapeutically useful field size is usually defined by the 90% or 95% isodose lines.

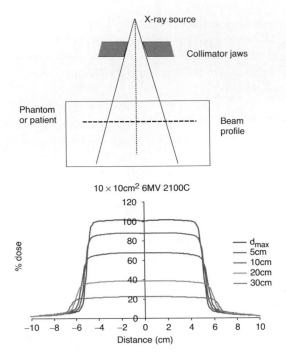

Fig. 8.6 Schematic illustrating the measurement conditions and a series of beam profiles taken at different depths in the phantom. This figure is reproduced in colour in the colour plate section.

8.4.1.2 Beam profiles and penumbra

A beam profile is a plot of dose across the beam in a direction perpendicular to the central axis, passing through the central axis, and normalized to the dose at the central axis.

There are several key features of the beam profile that can be demonstrated by looking at a profile measured at 10cm deep in water at the isocentre (Fig. 8.6). The beam profile is normalized to the intensity in the centre of the beam. Two important measurements that can be made from the profile are the flatness and the symmetry. The flatness is an expression of the difference between maximum dose and minimum dose across the beam profile at a defined depth in water in the central 80% of the beam. Beam flatness is characterized by a filter in a linac head (see Chapter 11). Beam symmetry is an expression of the dose at two points on a beam profile, each equidistant from the central axis. A beam is generally considered symmetric when these points are within 3% of each other but modern linacs usually produce beams well within this value. A flat and symmetric beam is desirable for radiotherapy planning; changes in flatness and symmetry may indicate changes in the beam energy or the steering of the beam in the wave guide. The flatness of the profile changes as a function of depth, field size and beam energy.

The edge of the beam is known as the penumbra. This can be defined as the distance between the 80% and 20% dose at the isocentric plane at depth. The penumbra is due to: the finite size of the X-ray source (Co-60 source, linear accelerator focal spot size) and scattered radiation; photons from the field and secondary electrons (released by photon interactions) out of the beam. The shape is affected by the focal spot size and shape

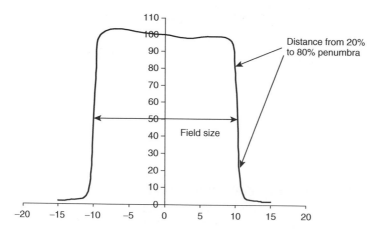

Fig. 8.7 Beam profile measured at 10cm deep for a 20cm by 20cm field size.

(which can be elliptical, rather than circular); scattering of photons and electrons; and the shape and properties of the collimators. The penumbra is larger for the inner jaws than for the outer jaws. The extent of the penumbra changes with depth in the medium. Looking further outside the penumbra the dose tails off and is mainly due to scattered radiation from the open part of the beam and transmission though the jaws of the treatment unit.

Due to divergence, beam profiles become wider at deeper depths. In addition, there is more scattered radiation so the penumbra becomes wider (Fig. 8.6). As the beam is attenuated the intensity is reduced as indicated by the fall off in the depth dose curve.

8.4.2 Isodose lines

A line joining all points in a plane that have the same percentage dose value is called **an isodose line**.

A chart showing a selection of isodose lines (usually in increments of 10% for a single beam) in any given plane is termed an isodose distribution. For superposed beams, fewer lines may be displayed for clarity.

8.5 Beam modifying devices

The treatment beam can be both shaped geometrically and can have its intensity distribution modified. These can be achieved using intensity modulated radiation therapy (IMRT) techniques. Another widely used method for achieving simple changes in intensity is to use **wedges**, while field shaping can be achieved using shielding blocks placed on an accessory tray.

8.5.1 Principles of wedges

Wedged beams are used for three main purposes:
  i) Combining beams from non-orthogonal angles,
 ii) Compensating for changes in surface shape,
iii) Compensating for changes in depth dose fall off for beams incident perpendicular to the wedged beam.

There are three types of systems that produce wedged beams:

i) Manual fixed physical wedges,

ii) Universal physical wedge,

iii) Dynamic wedges.

8.5.1.1 Manual or physical wedges

These are wedge shaped pieces of aluminium, brass or steel. A series of different wedges is usually in use e.g. 15, 30, 45, 60 degrees. These will have different physical dimensions and may be constructed of different materials. Physical wedges are less commonly used nowadays. This is because they need to be physically inserted in the position of the accessory tray. Inserting and removing the wedge can be difficult especially at non-zero gantry and collimator angles. Also carrying and storing them requires careful ergonomic design within the treatment room. They also block the light field used for setting up the patient, so the wedge is often inserted after the patient is set up, exacerbating the manual handling problems.

8.5.1.2 Universal physical wedge

A single physical wedge can be used to create a range of wedge angles by combining the wedge field with a plain or open field irradiation. This design of wedge is used in Elekta accelerators. The wedge is automatically positioned in the beam within the treatment head above the position of the mirror and below the monitor chamber. However, wedging can only occur in one direction which becomes a problem if the wedge is combined with a MLC, when being able to wedge both in the direction of the leaf movement and perpendicular to this may be useful.

8.5.1.3 Dynamic or virtual wedges

Dynamic or virtual wedges are created by moving a secondary collimator jaw across the treatment field while the beam is on. The amount of wedging is determined by the length of time the jaw is in the treatment field. Wedging in different directions can be achieved by movement of different jaws. For Varian dynamic wedges, the wedge factor is strongly dependent on field size and this effect needs to be carefully modelled within the treatment planning system. Off-axis and half-blocked wedged fields are created in the same way. Again care is needed in modelling the wedge factor.

8.5.2 Wedge factor (WF)

To deliver the same radiation dose to a point within a wedged field as for a plain field the number of monitor units set on the accelerator needs to be increased. This is achieved by use of a WF. The WF is defined as:

$$\text{Wedge Factor}\,(WF) = \frac{\text{Wedged Field}}{\text{Plain Field}}$$

The reciprocal of the WF indicates the increase in monitor units (MUs) required to deliver the same dose as for an identical plain field. The WF is a function of wedge type, wedge angle, beam energy, field size and shape, off-axis position and depth.

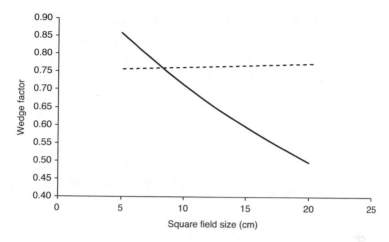

Fig. 8.8 Wedge factor variation with square field size for a manual wedge (dashed line) and a Varian dynamic wedge (solid line) at 15MV.

8.5.2.1 WF variation with field size

For physical wedges, the WF increases with field size. This is due to the amount of scattered radiation from the wedge increasing as the field size increases, and is typically of the order of 5–10% for clinically useful field sizes. For dynamic wedges, the variation with field size can be much larger, up to 50% with increasing field size.

8.5.2.2 WF variation off-axis

Off-axis, the WF for physical wedges tends to follow the profile of the wedge i.e. they increase or decrease in the wedged direction and remain approximately constant in the non-wedged direction. For dynamic wedges, the WF approximately matches the on-axis factor for the same field size with only small differences being observed.

8.5.2.3 Effect on %dd

For physical wedges, the wedge hardens the beam i.e. increases the mean energy of the beam making the %dd for a wedged field more penetrating than for a plain field because the lower energy components of the photon spectrum are absorbed in the wedge. This is more pronounced at lower energies (4–6MV) where Compton scattering is the predominant. At higher energies (>15MV), where pair production becomes more important, beam softening (a decrease of the depth dose with respect to the open field) is also possible. For dynamic wedges, there is no hardening of the beam and hence no change in the %dd or WF with changes in depth. The moving jaw attenuates the beams almost completely so there is no transmitted beam to harden.

8.5.3 Wedge angle

The wedge angle is defined as the slope of the line joining two points equidistant from the central axis and half the width of the field apart on the isodose curve which passes

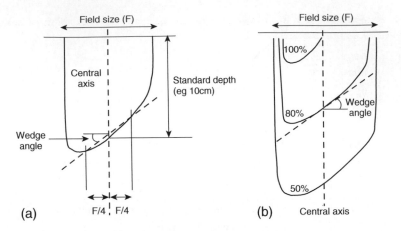

Fig. 8.9 a) IEC definition of wedge angle; b) alternate definition also in use.

through the central axis at a reference depth (usually 10cm). Alternate definitions include the angle between the tangent to a nominal isodose curve at the central axis, such as the 80% isodose curve, and the line perpendicular to the central axis (Fig. 8.9). Wedge angles between 10 and 60 degrees are clinically used.

8.5.4 Accessory trays

Even with the widespread use of MLCs it is sometimes necessary to use shielding blocks to shape the treatment field. These shielding blocks are located externally to the treatment head on accessory trays. The blocks may be fixed or free standing. The trays are usually constructed of Perspex. Double trays, where the shielding block is inserted between two layers can also be used. The single or double tray will attenuate the radiation beam and this effect needs to be accounted for in the monitor unit calculations by use of a tray factor defined as:

$$Tray\ Factor = \frac{With\ Tray}{Without\ Tray}$$

Lower energy megavoltage beams will be attenuated more by the presence of the tray. The tray factor increases a little with increasing field size, is due to increased scatter radiation from the tray. The scattered radiation also contains contamination electrons which increase the surface dose and hence reduce skin sparing when the tray is in use.

8.6 Shaped fields

This section considers the effects of shaped radiation fields. To understand how shaping a treatment field can affect the dosimetric characteristics of the beam it is helpful to consider the different components that contribute to the radiation dose for a radiation field.

8.6.1 Primary dose

This is the dose due to radiation incident on the phantom or patient (not scattered radiation from within the patient). It has two components: direct primary radiation and radiation scattered from within the treatment head (head scatter or sometimes termed collimator scatter).

8.6.1.1 Direct primary radiation

This component is radiation that has originated in the X-ray target and passes through the flattening filter without interacting. It is not dependent on field size.

8.6.1.2 Head scatter

This component is predominantly from radiation scattered from the flattening filter and also from the collimator jaws, primary collimator, monitor chamber and mirror. It is usually modelled as an extra focal source located at a scatter plane within the treatment head as shown in Fig. 8.10a. This component is field size dependent; as the jaw size increases more of the scatter source is visible and hence more radiation emerges from the treatment head.

8.6.1.3 Collimator exchange

The head scatter component is responsible for the effect in which the output for an elongated field (e.g. 30cm by 4cm) is different from the output for the same elongated field with the longer field dimension changed between the jaws e.g. 4cm by 30cm. This effect can be several percent and is due to the different jaws exposing different amounts of the scatter plane (Fig. 8.10b). Even though the area of the extra focal source exposed by the two elongated fields remains the same, the shape is changed resulting in the different amounts of head scattered radiation.

8.6.2 Scatter in the patient

In the patient, radiation is scattered from the irradiated volume. This scattered radiation component is called the 'phantom scatter' component. The larger the surface field size the more scattered radiation there will be (Fig. 8.11). The amount and proportion of scattered radiation will increase with increasing depth. Generally there is more scattered radiation at lower energies.

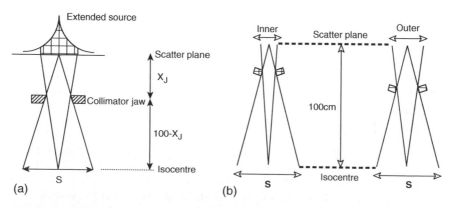

Fig. 8.10 Schematic illustrating how a) the head scatter component is represented by an extended source and b) the location of the jaws within the treatment head affects the size of the extended source visible from the isocentre.

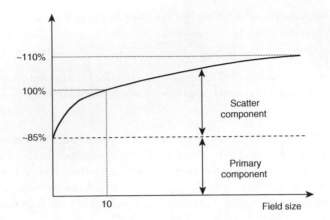

Fig. 8.11 Schematic diagram showing the increase in output with increasing field size and how the relative contributions vary. Nominal values are shown which are indicative only.

8.6.3 Methods to account for scattered radiation

There are a number of ways in which changes in output with field size and shape due to scattered radiation are considered. The following deals with some more common methods.

8.6.3.1 Scatter factor (SF)

Scatter factor (SF) is defined as the ratio of the total absorbed dose at a point to the primary dose at that point:

$$SF = \frac{\text{Total dose}}{\text{Primary dose}}$$

SF depends on beam energy, field size and depth. SF goes to 1 as the field size goes to zero.

8.6.3.2 Peak scatter factor (PSF)

This is a special case of the scatter factor where the reference point is on the beam axis at the depth of maximum dose.

8.6.3.3 Backscatter factor (BSF)

This is a special case of the scatter factor where the point of interest is at the surface of the phantom. It is used at energies below 400kV.

8.6.3.4 Output factor

The output factor is the dose at d_{max} for a set field size, normalized to a reference field size (usually a 10cm by 10cm field).

8.6.3.5 Tissue phantom ratio (TPR), Tissue maximum ration (TMR), tissue air ratio (TAR)

These quantities all give 'tissue dose' as a ratio of some reference dose which has been measured under reference conditions. To calculate the tissue dose you multiply the reference dose by the appropriate ratio. All these ratios refer to dose at the same point in the beam (i.e. at a fixed distance from the source), usually the isocentre. Hence there is no divergence and no SSD dependence. These ratios depend on depth (d), energy (E) and size and shape of the field. The depth dependence is due to attenuation and scattering and does not incorporate a divergence effect.

8.6.3.5.1 Tissue phantom ratio (Fig. 8.2)
The TPR is defined as the ratio of the absorbed dose at a point on the central axis at any given depth to the absorbed dose on the central axis at the same distance from the source but with the surface of the phantom moved so that the point is at a specified reference depth. The collimator settings remain unchanged.

8.6.3.5.2 Tissue maximum ratio
TMR is a special case of the TPR in which the reference depth is the depth of maximum dose.

8.6.3.5.3 Tissue air ratio
A TAR is defined as the product of the tissue maximum ratio (TMR) and the peak scatter factor (PSF). TARs were used for Co-60 units but have been replaced by TPRs for megavoltage calculations. TAR was originally defined as the ratio of the absorbed dose at a point on the central axis at a depth in tissue, to the tissue dose, in air at the same point in the beam. This leads to problems at higher energies where significant thicknesses of build up material are required for electronic equilibrium and hence dose in air is not being measured. At high energies large build up thickness, as are required, give rise to attenuation and scattering; therefore the true primary dose is not being measured.

8.6.3.6 Equivalent square field
The depth dose characteristics of rectangular and circular fields can be represented by calculating an equivalent square field size with the same characteristics. This is not a square field of the same area but of a field whose dimensions can be represented by:

$$\sigma = \frac{2ab}{(a+b)}$$

where a and b are the two sides of the rectangle and σ is the dimension of one of the sides of the equivalent square. This is only an approximation. The BJR Supplement 25 has extensive tables of equivalent field sizes.

8.6.3.7 Scatter from irregularly shaped fields
For irregularly shaped fields, the Clarkson sector integration technique is often employed. In this the irregular field is divided into a series of sectors of circular fields. These sectors can then be summated to determine the output from the irregular field (see section 10.3.10). Dose calculations for irregularly shaped fields are usually undertaken within the treatment planning system.

Chapter 9

Electron beam physics

G Pitchford and A Nisbet

9.1 **Electron beams used in clinical practice**

High energy electron beams may be used in preference to megavoltage photons in situations where their physical dose distribution offers advantages over those of photons. The shape of an electron beam depth dose curve is characterized by a small skin sparing effect, a relatively uniform dose for a definite depth around the depth of maximum dose (d_{max}) and a relatively steep fall off with a finite range. A representative set of central axis depth doses can be seen in Fig. 9.1.

The primary aim of electron beam therapy is to offer a method of treating target volumes situated on the surface of a patient, or extending below the surface to a limited depth. This can be useful where underlying tissue with a higher Z number, such as cartilage or bone, results in a higher absorbed dose when kilovoltage X-rays are used (due to an increase in photoelectric interactions).

Although microdosimetric properties (i.e. linear energy transfer distributions) may be different for photons versus electrons, suggesting different radiobiological effectiveness (RBE), cell survival data has indicated no difference in their RBE values.

Historically electron beams were produced predominantly in the energy range 5–35 MeV; however there is little advantage for electron beams compared to high energy photons at the higher electron energies (see Section 9.2). Modern linear accelerators now provide high energy electrons in the energy range 4–20 MeV.

The middle energies are used more frequently with typical usages:

8MeV–12 MeV	70%
6MeV & 14–16MeV	23%
4MeV & 18–20MeV	7%

Electrons are used in a number of clinical settings:

- Skin and lips,
- Chest wall and neck, both after surgery and for recurrent disease,
- Boost doses to limited volumes.
    1. Tumour bed for breast—assumed GTV,
    2. Posterior neck nodes overlying spinal cord—partial CTV,
    3. Scar areas.
- Combined electron and photon beams e.g. parotid,
- Total skin irradiation for mycosis fungoides and cutaneous lymphomas.

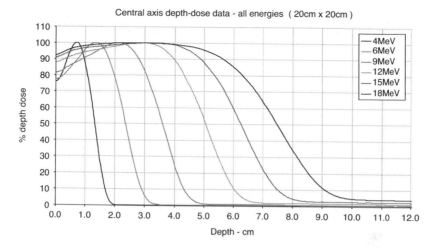

Fig. 9.1 Central axis depth doses, 4 MeV–18 MeV, 20cm x 20cm applicator. This figure is reproduced in colour in the colour plate section.

Although a full literature review of the use of electrons in these clinical settings is beyond the scope of this book, the following overview provides an indication of their efficacy in appropriate situations.

Electrons have been used for many years to treat cutaneous basal and squamous cell carcinomas. Electrons have also been used with success in non-melanoma skin cancers and have been shown to be an efficient treatment in the management of Kaposi's sarcoma.

Post mastectomy patients can be treated with low energy electron beam radiation therapy, protecting underlying lung without sacrificing local disease control. Electron beams have been used extensively in head and neck cancer, particularly when treating tissue overlying the spinal cord. Once photons have treated the region to a dose close to spinal cord tolerance, an electron field can be used to boost the dose in this region, whilst continuing to treat tissues away from the cord with photons. The relatively rapid fall off in dose at depth helps keep the dose to the spinal cord below tolerance. The electron and photon fields need to be carefully matched. This technique, however, will wane as intensity modulated radiotherapy makes it redundant.

Total skin electron therapy (TSET) is the most effective single agent for the treatment of mycosis fungoides, especially for patients who have thick generalized plaque or tumorous disease and this technique may be used selectively for extra cutaneous disease. There are well defined dose response relationships for achieving a complete response as well as the durability of this response. Electrons in the energy range 2–9MeV are predominantly used given the rapid drop off in depth dose and a low bremsstrahlung component. This enables superficial skin lesions to be treated to a depth of around 1 cm without exceeding bone marrow tolerance. A number of treatment techniques have been developed broadly based on either:

(1) A translational methodology where the patient lies horizontally and is moved relative to a beam of sufficient width to cover the transverse direction of the patient.

(2) A large field technique where a standing patient is treated with a combination of broad beams and extended SSDs.

Although some of the treatments traditionally carried out by electrons are now planned with IMRT/VMAT, electron beams are likely to continue to have a role within radiotherapy for the foreseeable future. Some groups are exploring the use of intensity and energy modulated electron beam radiotherapy which may be a viable treatment option for shallow head and neck tumours. The use of intensity modulated radiotherapy incorporating conformal electron irradiation has also been shown in retrospective planning studies to minimize heart dose in post mastectomy breast radiotherapy. Therefore future radiotherapy techniques involving electron beams cannot be discounted.

Most electron treatments are delivered as single fields with normal incidence at a fixed source-surface distance (SSD). In some instances the electron applicator is placed against the skin rather than with a fixed stand-off distance. It may also be necessary to treat at a non-standard SSD if the patient surface prevents positioning of the electron applicator at the standard SSD. Corrections may be applied to electron output for small changes in SSD. Strictly speaking such corrections should use a modified inverse square law but in practice, the application of a correction factor in terms of a few percent per centimeter will usually suffice.

9.2 Energy ranges

The interactions of electron beams are described in Chapter 4 and the concept of the continuous slowing down approximation (CSDA) introduced. The point at which the electron has lost all its energy identifies its range. If all electrons in the beam lose energy in the same way, all electrons will be stopped at the same depth. However, due to scattering, the actual depth reached will vary, this is known as range straggling. The path length (total distance travelled) will be the same for all electrons of the same energy (Fig 9.2). This explains the characteristic shape of an electron beam depth dose curve, with its high surface dose, relatively constant dose around the depth of maximum dose and then rapid drop in depth dose to a low dose tail arising from a bremsstrahlung X-ray component (Fig. 9.1). The electron range in cm can be approximated by dividing the mean electron energy in MeV by 2.

Depth doses in clinical data tables are usually presented in terms of the depth related to a specified percentage depth dose. When the oncologist has assessed the greatest depth of the relevant target volume then the energy required to effectively cover this volume can be read off the depth dose tables for that field size. Table 9.1 shows a selection of typical depth doses.

The incident energy, E_0, is the mean energy at the surface of the patient. The practical range occurs at the intersection of dose fall off and the bremsstrahlung tail at about the depth of the 1–5% dose level and represents the range of the incident electron beam.

Of interest in assessing target coverage is the surface dose (see Section 9.5), the depth of the maximum dose and the minimum tumour dose to be encompassed by the therapeutic range usually taken to be 90%, or occasionally 85%, isodose. In addition the dose beyond the target is of clinical relevance in assessing dose to underlying tissue.

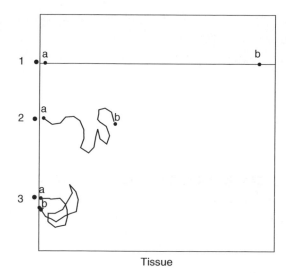

Fig. 9.2 Consider 3 electrons each travelling in tissue. They lose their energy, depositing it in tissue as dose, on their journey from A to B. The distance AB is the same for each, i.e. they have equal path lengths. However, the depth they reach in the tissue is very variable and depends on how scattered their path is. When applied to a huge number of electrons, the 'range' (depth in the tissue) will straggle over a set of distances; this is range straggle.

9.3 **Percentage depth dose**

Referring again to Fig. 9.1, the depth dose curves exhibit a number of properties:

- High surface dose which increases with energy (opposite to what happens for photon beams).
- Build up to the depth of maximum dose (which can be broad at high electron energies).
- Rapid dose fall off beyond dose maximum (higher energies give less rapid fall off).
- Low value bremsstrahlung tail which increases slightly for higher energy beams.

There are a number of simple rules of thumb relating to central axis depth dose values:

- The therapeutic range 90% to 85% (on the fall off side of the curve) $\sim \frac{1}{3} E_0$ cm.
- Mean energy decreases by 2 MeV per cm in water or soft tissue and hence the practical range is $0.5 E_0$ cm.
- The 50% depth dose lies half way between the therapeutic range depth and the practical range depth. However the depths of the 90% and 50% doses can be significantly reduced with small field sizes whilst the practical range remains unchanged (see Section 9.4).

Table 9.1 Representative depths for a particular field size/applicator

Incident Energy (MeV)	6	8	10	12	14	16
Depth of maximum dose (cm)	1.2	1.8	2.3	2.9	3.3	3.7
Depth of 90%	1.7	2.3	3.1	3.9	4.7	5.2
Depth of 80%	1.9	2.7	3.5	4.3	5.1	5.6
Depth of 50%	2.3	3.0	4.0	4.9	5.8	6.4
Depth of 10%	3.0	3.9	5.2	6.0	7.3	8.3

9.4 **Factors affecting depth dose: field size**

For moderate and large field sizes (>10 cm), there is little variation in central axis depth doses with field size for a particular energy. However, the build up region may differ because scatter from collimators and other structures in the linear accelerator head has a marked effect on the detail of the depth dose curve from the surface to d_{max} and this scatter will depend on the applicator/field size setting.

The little variation in central axis depth dose beyond d_{max} occurs when the distance from the central axis of the beam to the field edge exceeds the lateral range of scattered electrons setting up an electronic equilibrium. This is particularly true for electron energies up to 10MeV. This is demonstrated in Figs 9.3 and 9.4 which show respectively the variation of the depths of the 50% dose point (d_{50}) and d_{max} with field size and electron beam energy.

For energies above 14MeV (according to accelerator design) it may be found that the position of the maximum (100%) central axis dose, D_{max}, may be closer to the surface at the highest energy than a lower E_0. This can be seen in Figure 9.1 for the 18MeV curve.

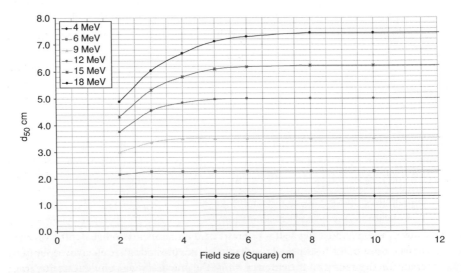

Fig. 9.3 Effect of field size on d_{50} as a function of beam energy. This figure is reproduced in colour in the colour plate section.

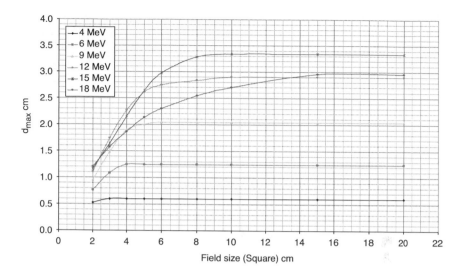

Fig. 9.4 Effect of field size on d_{max} as a function of beam energy. This figure is reproduced in colour in the colour plate section.

For small field sizes, less than 6 cm square, the situation can be complicated and careful consideration of depth dose and isodose data is recommended to ensure adequate clinical coverage. As the field dimensions decrease there is a loss of scattered electron equilibrium at the central axis of the field; this can occur even if only one of the dimensions falls below the practical range for the electron energy being used. The dose maximum and other high dose values are displaced towards the surface with a corresponding increase in the surface dose. The practical range, which is dependent on the value of E_0, remains unchanged and hence the fall off gradient is reduced bringing a reduction in the therapeutic range. This is demonstrated in Fig. 9.5.

This situation might suggest using a higher energy electron beam but the higher the value of E_0 the more pronounced the effect of dragging the therapeutic range depth towards the surface. There is also a lateral constriction of the high dose isodoses (see Section 9.6) that reduces the volume encompassed by the therapeutic dose value. These effects need to be considered when prescribing small field electron treatments.

9.5 Build up and skin sparing for electrons

The surface dose varies from approximately 75% to 95% depending on the initial electron energy. The values increase with increasing energy as can be seen by the representative values in Table 9.2. There is little variation with field size with the caveat that small field sizes at high energies may deviate from this general situation.

There is little skin sparing for the deeper tumours such as posterior neck nodes. However, when irradiating a skin lesion it may be necessary to use bolus to ensure a surface dose of 90%–100%. This may then require a higher electron energy to correct for the required therapeutic range. Tables may be provided for surface doses obtained by the use of standard thicknesses of bolus material for example 0.5cm, and 1cm.

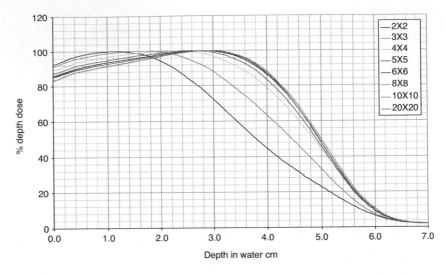

Fig. 9.5 Effect of field size on depth dose for 12 MeV. This figure is reproduced in colour in the colour plate section.

The bolus may be in the form of a flexible tissue equivalent material such as Superflab™ which comes in selected thicknesses and usually covers the entire electron field and is used to increase the surface dose. This may also be achieved by using a plastic sheet attached to the applicator or end frame. Wax can also be used as bolus and is often customized for each patient's treatment. It may be used to reduce the penetration of the beam in a particular part of the field to perhaps protect a vital organ, or to flatten out irregular surfaces such as those found in the head and neck region.

9.6 Isodose curves for electrons

Electron isodose curves are very different from the geometrical shape of megavoltage photon beam. Since the electron beam energy is being constantly degraded as it penetrates through the patient there is an increasing amount of laterally scattered

Table 9.2 Electron surface doses, 6 MeV–18MeV, 6cm–25cm applicators

Energy (MeV)	Electron Surface Doses Relative to 100% at Maximum Dose				
	Applicator				
	6 cm	10 cm	15 cm	20 cm	25 cm
5	74.8	75.1	75.5	76.0	75.7
9	77.6	77.9	78.2	79.0	78.7
12	84.1	83.4	83.6	84.6	84.3
15	88.4	88.0	87.8	88.4	88.4
18	91.8	91.4	90.7	90.7	90.7

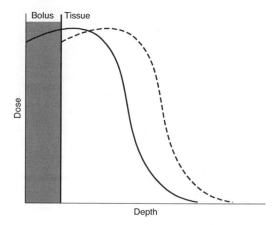

Fig. 9.6 PDD curves for an electron beam with and without bolus.

electrons which results in an increasing penumbra with depth. As the electron beam passes through the patient it expands just past the surface of the patient, until a depth of approximately 70% is reached when it starts to narrow. Close to the surface the 50% isodose line follows the geometric field edge. The higher dose levels increase their separation from the 50% line and hence there is a constriction of the width of the high dose values which means that the high dose volume, above 90% is narrower at the therapeutic range depth than at the surface or D_{max}. These features can be seen in Fig. 9.7. The constriction is approximately $2E_0$ mm on each side of the field and a larger value across a diagonal of a rectangular field. The value is accelerator dependent.

Lower dose levels bulge outwards so that at a depth there can be significant doses delivered outside the geometric edge of the field. Hence clinical consideration needs to

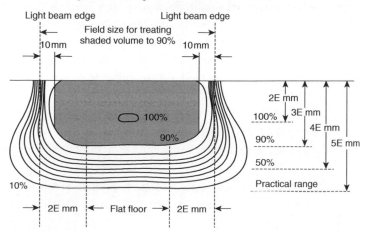

Fig. 9.7 Electron isodose curve. From *Handbook of Radiotherapy Physics: Theory and Practice* edited by P Mayles, A Nahum, J C Rosenwald, Copyright 2007. Reproduced with permission of Taylor & Francis Group LLC.

be given to the width of the beam at the therapeutic range and also to the closeness of any critical organs to the geometric edge of the electron beam.

9.6.1 Clinical aspects

If the target volume is superficial then the energy selected to encompass the therapeutic range may not produce a sufficient surface dose. In this case a higher energy needs to be selected with bolus or an energy degrader to increase the surface dose and maintain the required dose to the distal target volume. This is usually successful with only a minor reduction in the fall off gradient.

An optimum field size may have to be chosen if there is a conflict between the need to have a wider beam at the patient's surface to encompass the constricted 90% isodose at a depth and the proximity of a critical organ to the lateral edge of the target volume. This can be a particular problem with small volumes around the nose and cheek, and close to the eyes. In this situation the use of small irregular shaped fields are produced by the use of cut-outs manufactured from lead sheets or low melting point alloys; clinical data may need to be verified by dose measurements obtained using the patient specific cut-outs.

If the thickness of shielding is too thin the dose on the surface of the patient, instead of being reduced, may be enhanced due to the production of forward scattered electrons and bremsstrahlung from the high atomic material on the skin surface (much like X-ray production in a linac target). As a rule of thumb the minimum thickness of lead required in millimetres is approximately half the mean surface energy of the beam in MeV. For low melting point alloy (LMPA) this should be increased by a factor of 1.2. Generally this will result in a transmission of less than 5% at lower energies to 10% at 20MeV.

The ability to join two or more adjacent fields can be useful in a number of clinical situations:

♦ Adjacent areas having received prior irradiation e.g. large field head and neck fields followed by photon field to a primary and matched electron fields to posterior neck nodes to spare further irradiation of the spinal cord.

♦ To treat an irregular surface e.g. around a chest wall, scalp. This may mean employing fields angled towards each other or at right angles, potential overlaps can be mitigated by the use of absorbers at the field edges.

♦ Varying depths across the target volume.

♦ A larger area than a standard applicator e.g. skin lymphoma.

However they do present significant problems due to the shape of the electron isodoses:

♦ A match at the surface edge of abutting fields gives hot spots at a depth,

♦ A match at a depth by leaving a calculated gap at the surface produces a cold spot within the target volume,

♦ Significant differences in hot or cold areas can be produced by small variations in the relative positions of the beams,

These problems can be alleviated by employing a number of strategies

◆ Position abutting edges away from any critical area,

◆ Use moving junctions, similar to some photon techniques e.g. craniospinal irradiation, to 'smudge out' the overlap area,

◆ Slightly angle beams away from each other to reduce hot spot area,

◆ Utilize strips of absorbers or energy degraders along the match line,

◆ Doses and dose distributions should be tested experimentally prior to treatment and perhaps also during treatment in the more complicated circumstances,

◆ Arc therapy may be used instead of adjacent fields but the implementation of this type of technique requires significant technical effort.

9.7 Effects of surface obliquity and inhomogeneities on dose distributions

9.7.1 Surface obliquity

Oblique incidence occurs where there is stand-off from the patient's surface to part of the electron beam. A common example of this is treatment of the whole of the chest wall.

If we define an angle α as the angle between the central axis of the beam and the normal to the patient's skin surface then for:

◆ α less than 20° there is little effect on the depth doses and the isodose curves follow the skin surface,

◆ α between 20° and 30° the isodose curves still follow the surface but if there is stand-off at edge of the collimated area then the penumbral region is widened,

◆ α between 40° and 60° the depth dose is reduced and the surface dose is increased; the increase in surface dose may be greater than the loss of fluence due to the accompanying extended SSD producing hot spots,

◆ α greater than 60° the percentage depth dose no longer has its characteristic shape (more akin to superficial X-rays) the value of the practical range changes and there is a steep increase in the maximum dose.

It is advisable if possible to angle the field so that the fall off around the collimator edges are equalized which reduces the more serious effects of surface obliquity. If this is not feasible then the difference could be compensated for by the use of variable thickness bolus and possibly a higher electron energy.

9.7.2 Inhomogeneities

Inhomogeneities have an effect on the dose distribution which are dependent on:

◆ Energy and field size of the electron beam,

◆ Size of inhomogeneity relative to the field size,

◆ Its shape and composition.

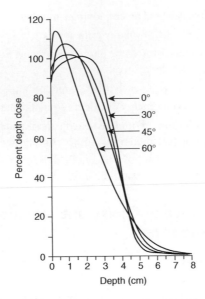

Fig. 9.8 The variation in depth dose with variation of angle of incidence along the central axis for 10 MeV electrons: note how the increased angle of incidence beyond 30 degrees leads to a decrease in the depth of the maximum dose, but greater than 60 degrees leads to a large increase in the maximum dose. Reproduced from Hoskin and Coyle, *Radiotherapy in Practice: Brachytherapy, Second Edition*, 2011, with permission of Oxford University Press.

There are two main effects:

1. Absorption and the associated shift in the depth of isodose values; usually most noticeable within a large inhomogeneity and beyond its limit e.g. lung beneath a chest wall.

2. Scatter differences between materials; these can be significant for small homogeneities or in the interface region with a large homogeneity.

The scatter effects are complex with a higher proportion of electrons scattered from a high density region into a low density region than vice versa. This gives a low dose area within the high density region or distal to it and conversely a high dose area within a low density region and beyond it.

The effect is energy dependent, increasing with increasing energy requiring an estimation of the effective electron energy at the depth of the inhomogeneity. A Monte Carlo based treatment planning system (section 10.6.4.4) would give the best approximation of these interface effects which can be founding a number of situations:

- Interfaces involving air, lung and bone,
- Surface irregularities such as shaped bolus, lead cut-outs, nose, ear.

The ear can present significant problems which can be tackled by the use of a tissue equivalent plug e.g. wax to circumvent the air/tissue interface.

9.8 **Internal shielding**

Internal shielding is used to spare structures lying below the target volume and is usually encountered in the irradiation of lips, cheek, ear and eyelids. There is a need to calculate the minimum thickness of shielding material to give the required shielding which is energy dependent. It may be that the there will be a restriction on this thickness due to the available space e.g. under an eyelid.

Any shielding material will produce backscatter and this will increase the dose at the tissue/shield interface. The excess dose increases with increasing atomic number, Z, of the shielding material and decreasing electron energy and can be significant, with increases over 50% being possible. The use of low Z material as a coating mechanism can significantly reduce this excess dose by absorption of the low energy backscattered electrons. Dental wax is frequently used for this purpose. Typically 2mm of lead may be used for shielding with a coating of 8mm of wax. This may be feasible in the case of the inside of a lip. However any shielding may be impractical in the case of the inner surface of an eyelid. The use of internal shielding may lead to a compromise in the prescribed dose.

Chapter 10

Radiotherapy treatment planning

N MacDougall, C Nalder and A Morgan

10.1 Treatment planning

In general, the term treatment planning refers to all the decisions and processes required to design a patient's radiotherapy treatment. The complexity of the process depends on the treatment intent and technique but the design of most modern treatments has many basic elements in common.

The target site will determine the patient's treatment position and any immobilization applied. Imaging of the patient in the chosen position then allows the target volume to be defined and may also provide physical data on which to base dosimetry calculations. The dosimetry calculations vary in complexity from the determination of dose at a single point within the patient to a full 3D dose calculation summarized by dose volume histograms for target volumes and organs at risk (Section 10.3.6). Dose calculation models employed on a treatment planning systems (TPS) are usually called algorithms. There may be more than one type of algorithm available on a TPS (Section 10.4.4).

In all cases the output of the process will be the treatment unit parameters required to execute the planned treatment together with a set of instructions for aligning the patient on the treatment unit. Some reference imaging should also be produced to permit the patient set up on the treatment unit to be verified.

10.1.1 Patient position and immobilization

Patient position should be chosen carefully and must not change between the planning and treatment stage. It is important that the position is optimal for dosimetry and that the patient remains relaxed and comfortable if stability is to be maintained throughout the treatment course. Other constraints on patient position may be imposed by the aperture size of scanners or by the location of the therapy unit head during treatment delivery.

Immobilization of the patient in the treatment position has been shown to significantly reduce setup errors. Such errors will result in a failure to deliver the planned treatment resulting in a possible under dose to the target or overdose to nearby organs at risk. Various immobilization systems are used depending upon the treatment site. Immobilization may be minimal, as in the use of a knee rest and ankle stocks for pelvic radiotherapy, or more complex and customized for the individual patient, as in the case of thermo-plastic shells for head and neck treatments or relocatable stereotactic frames for brain. Effective systems must be well tolerated by patients and practical to use with short set up times.

It is important that each immobilization system is evaluated in terms of the accuracy and reproducibility of patient positioning. This is most commonly achieved by a comparison of the reference images with treatment verification images in terms of the position of bony landmarks. Any discrepancy observed between the two will be due largely to patient set up error.

10.1.2 Organ motion

Immobilization systems assist in setting up the patient reproducibly but they cannot prevent the internal motion of the target volume and other organs with respect to the patient's bony anatomy. These internal physiological motions cannot be easily controlled and if not handled correctly will also lead to inadequate treatment delivery. Examples include respiratory motion and the movement of the prostate due to bladder and rectal filling.

Techniques are being developed to manage organ motion, the approach used depending on the nature of the movement. Predictable, regular motion such as that resulting from steady respiration can lend itself to gated delivery whereas slower more unpredictable motion, such as that of the prostate, requires soft tissue imaging with online set up correction prior to treatment delivery. If techniques are not available to manage internal organ motion it is necessary to estimate its extent for each site and increase the target volume accordingly (Section 10.3.6).

10.1.3 Tumour localization and the patient model for dosimetry

Tumour localization refers to the identification of the target volume relative to the patient's internal anatomy and external skin surface. The process relies on the acquisition of suitable images of the patient in the treatment position.

In conventional 2D planning techniques, the imaging consists of a pair of planar radiographs acquired on a radiotherapy simulator. Target localization is carried out with respect to bony anatomy and radio-opaque markers with minimal shielding applied. For 3D conformal radiotherapy, the tissues to be treated are imaged and identified as the target. The treatment fields are designed to conform to the target plus an additional safety margin and this leads to a more customized treatment. The standard imaging modality is CT scanning, providing soft tissue information which can be enhanced further by the use of injected contrast media. Registration of the CT planning scans with other imaging modalities such as MRI or PET-CT extends further the information available to the clinician on which to base the target localization. Volumes are most often outlined on transverse slices through the patient although many planning systems now have the facility to reconstruct images in any plane.

For alignment of the patient during treatment, anterior and lateral reference points on the skin surface are needed. A common practice is to use small skin tattoos near the treatment site, positioned if possible at points where the skin is relatively stable. In order to relate the position of surface tattoos to that of the target volume, radio-opaque markers are placed over the tattoos so that their position is recorded on CT scans or radiographs. If the patient is in an immobilization device, the reference points can be marked on this rather than on the patient's skin. The set up instructions provided as part of the planning process locate the isocentre with respect to the surface reference points. However, accuracy of set up must always be confirmed with verification imaging.

Dosimetry in conventional 2D planning is performed on a single transverse outline of the patient. The target volume is reconstructed within the outline by demagnifying the target dimensions defined on the localization radiographs. Measurements are made from reference positions marked on the film and the outline. Organs at risk and tissue inhomogeneities are reconstructed in the same way and standard values for tissue electron densities applied to the structures. Optimization of the dose distribution is restricted to this single plane so it is important that the plane is representative of the shape of the patient throughout the target.

The use of CT imaging has been essential in the move from 2D to 3D treatment planning and dosimetry. In addition to its role for target localization, the CT image also provides an accurate 3D physical model of the patient. This includes both the external surface contour and the position and density of tissue inhomogeneities. In modern treatment planning systems (TPSs), the CT image forms the basis for the dose calculation as X-ray attenuation data (Hounsfield units) may be related directly to tissue electron density. Dose homogeneity can now be optimized in 3D and this is achieved in practice by a variety of techniques. Wedges may be applied in any direction rather than being restricted to a single plane, additional sub-fields or segments may be used to boost areas of low dose or full inverse planned IMRT may be used.

10.2 The simulator and the CT scanner

10.2.1 Structure and use of a simulator

A simulator is a machine designed to mimic the mechanical movement and beam geometry of treatment machines: linear accelerators (linacs) and cobalt units. As such it has a gantry, collimator and treatment table that can all rotate and move as a treatment machine. It looks like a very slim linac! However, a simulator's radiation source is a diagnostic energy X-ray source (kV) and a diagnostic X-ray detector replaces the linac portal imager. The simulator head (where the X-ray source is) can move in and out to vary the source to axis distance (SAD). This can be helpful for mimicking cobalt units (commonly have SAD of 80cm) as well as linacs (SAD 100cm).

The main purpose of a simulator is to use the kV X-ray source to image patients and plan treatments. The classic example of a simulator planned treatment is an X-ray image of a patient's bony anatomy with a chinagraph pencil marking the field edges and any shielding necessary. This process requires the patient to lie on the simulator couch for a time whilst the treatment borders are decided upon. The simulator does not capture any patient contour or electron density information, so it does not acquire all the information we require to plan treatment. The patient shape contour can be obtained with a piece of flexible wire, but there is no way to find out anatomical electron density information.

10.2.2 Use of a CT scanner in radiotherapy planning

CT scanners capture patient anatomical information (shape, size etc.) and electron density information. The data contained in a CT scan allows patient treatments to be

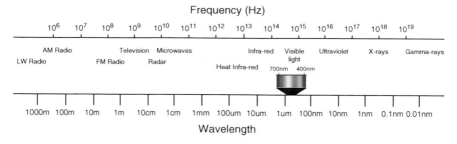

Fig. 2.9 The electromagnetic spectrum.

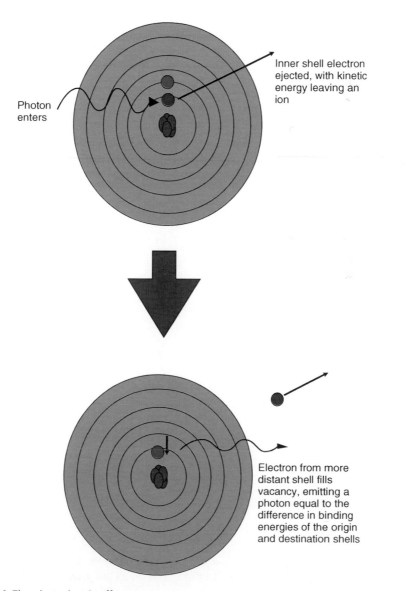

Fig. 3.4 The photoelectric effect.

Fig. 4.3 Nuclear reactor rods in a water filled cooling pond producing Cerenkov radiation. Image courtesy of Nordion Inc.

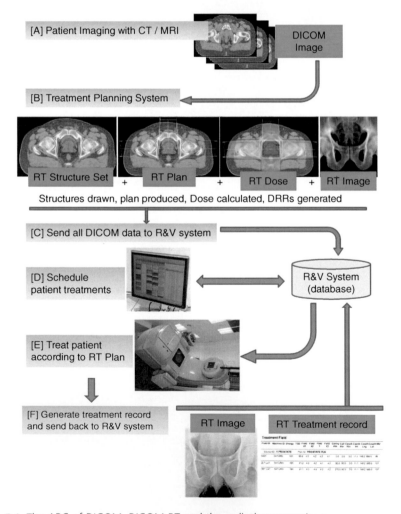

Fig. 5.1 The ABC of DICOM. DICOM-RT and the radiotherapy pathway.

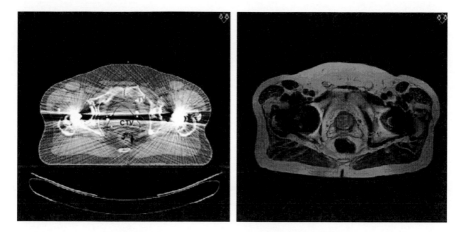

Fig. 6.1 The use of CT-MR image registration to define target volumes in pelvic radiotherapy in the presence of bilateral hip replacements.
Charnley, N., *et al.* (2005). The use of CT-MR image registration to define target volumes in pelvic radiotherapy in the presence of bilateral hip replacements. *British Journal of Radiology* **78**: 634–636.

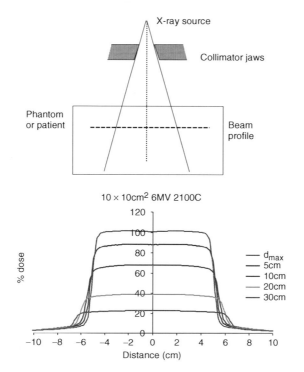

Fig. 8.6 Schematic illustrating the measurement conditions and a series of beam profiles taken at different depths in the phantom.

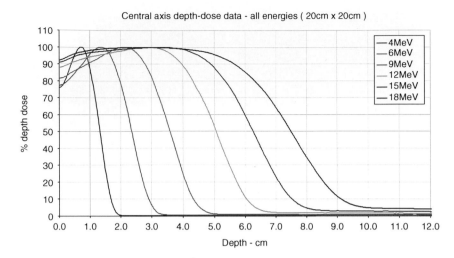

Fig. 9.1 Central axis depth doses, 4 MeV–18 MeV, 20cm x 20cm applicator.

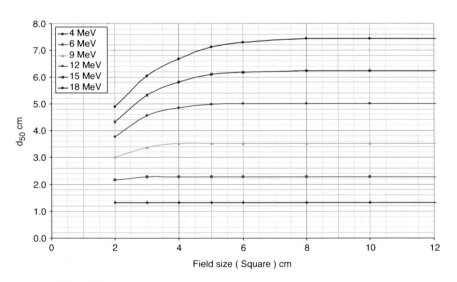

Fig. 9.3 Effect of field size on d_{50} as a function of beam energy.

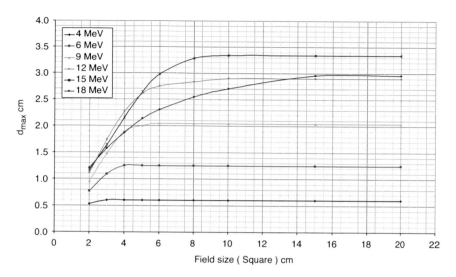

Fig. 9.4 Effect of field size on d_{max} as a function of beam energy.

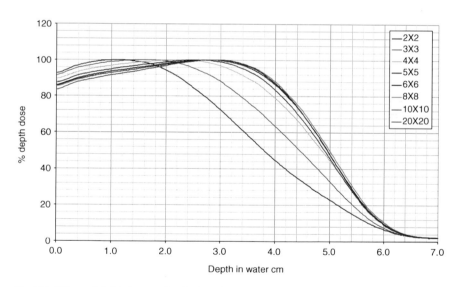

Fig. 9.5 Effect of field size on depth dose for 12 MeV.

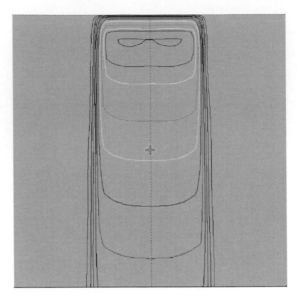

Fig. 10.1 Isodose distribution from a 6MV photon beam incident on water at 100cm FSD (100% is the red line).

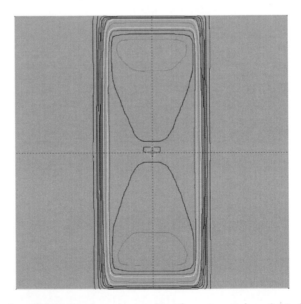

Fig. 10.2 The dose distribution in water resulting from an evenly weighted parallel pair. The beam energy is 6MV and the phantom width 16cm.

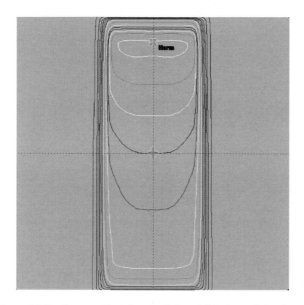

Fig. 10.3 Isodose distribution resulting from a 2:1 weighted parallel pair.

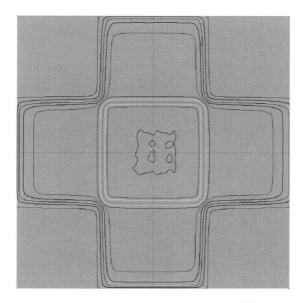

Fig. 10.4 Isodose distribution resulting from a four field brick field arrangement.

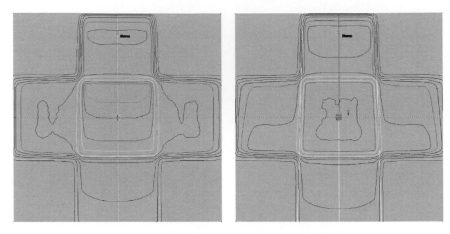

Fig. 10.5 Isodose distribution resulting from three plane fields at right angles (left) and for the same set up with 45 degree wedges added to lateral fields (right).

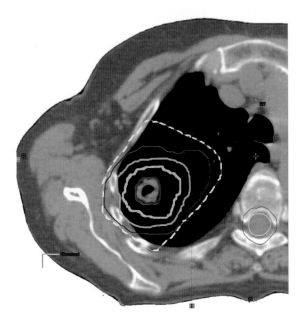

Fig. 10.6 Lung tumour showing GTV (blue), CTV (cyan), ITV (green), PTV (red), treated volume (yellow dotted line), OAR (spinal cord, yellow line), PRV (magenta).

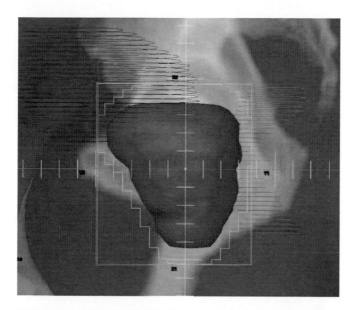

Fig. 10.8 BEV showing PTV (red), bladder (blue) and MLC field edges (yellow).

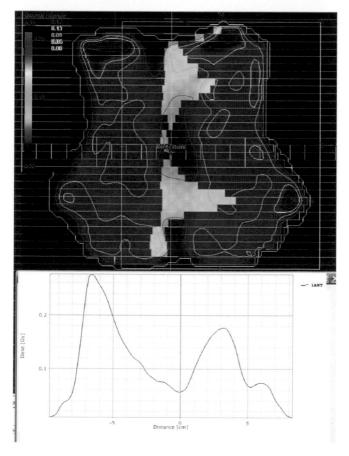

Fig. 10.11 BEV of MLC half way through an IMRT field. Dose profile taken through vertical axis.

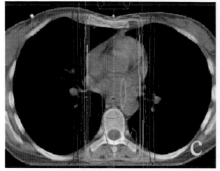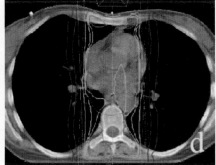

Fig. 10.14 The same patient treatment calculated with pencil beam, c, and superposition, d. The more accurate modelling of scatter in d can be seen by the low dose levels (blue lines) spreading laterally, but the high dose lines (green and orange) being drawn in medially. Reproduced from C. Irvine, A. Morgan, A. Crellin, A. Nisbet & I. Beange (2004). Clinical Implications of the Collapsed Cone Planning Algorithm, *Clinical Oncology*, **16**, 2: 148–154, with permission from The Royal College of Radiologists.

planned in computer software: virtual planning and virtual simulation. So, instead of having to have the patient lie on the simulator couch whilst X-ray images are taken and treatment borders decided upon, the patient can have a quick CT scan and leave the room. The CT scan contains all of the radiological information one requires to plan the treatment allowing the Virtual Simulation (V-Sim) process to take place 'off line'. This allows the patient to go home and the clinician to V-Sim patient treatments at a different time.

10.2.3 V-Sim software and approaches

The increase in sophistication of computers has allowed the development of virtual simulation V-Sim software. For V-Sim we take the patient CT scan and import it into the V-Sim software. The V-Sim software mimics the linac and allows the user to change all relevant linac parameters (gantry, collimator, couch angle; field size, MLC position etc.) and generate a 'virtual' treatment beam. As one changes the various parameters, the software will update beam's eve view (BEV) projections (see Fig. 10.9) and Digital Reconstructed Radiographs (DRRs) discussed in Section 10.4.1. This gives the user an image similar to that produced by the simulator: albeit a snapshot of the patient anatomy with no movement (see Fig. 10.8).

10.2.4 4D basic principle and definition

One of the weaknesses of V-Sim is that we are dealing with a CT scan which is effectively a snapshot of the patient. When treating moving anatomy (such as lung) we will not have captured all the information about tumour location (as the tumour is moving). When using a simulator one would have used fluoroscopy to image tumour movement and decide on treatment margins. When using CT, the closest we can get to this is to use 4DCT. What this means is that we take multiple CT scans of the patient in time (4D meaning the fourth dimension: time). 4DCT scanning is usually combined with 'gating' to capture the tumour in all parts of its movement cycle. An analogy of this is if you were bouncing a tennis ball off the floor and catching it. A photograph (CT scan) would only tell you where the ball was at the point in time when the photo was taken. To see the full range of where the ball goes we would have to record a video of the motion (4DCT).

10.2.5 The rise and fall of simulation

The rise of V-Sim has conversely seen a decrease in the need for a simulator and many departments now function without one.

10.3 Basic concepts and conventional treatment planning

10.3.1 Fixed focus to skin distance (FSD) planning and isocentric planning

In modern radiotherapy departments almost all treatments units are isocentric and this is the preferred approach to treatment planning and delivery. During an isocentric treatment, because the axis of gantry rotation is fixed within the target volume, the FSD to the patient will vary as the gantry rotates. Isocentric delivery allows the beam to be repositioned around the patient with the minimum of intervention. In contrast, fixed FSD treatments apply a standard FSD for each treatment field and as a result, multiple beam fixed FSD treatments typically require large lateral and vertical couch movements in order to set each new beam entry point. Even so, the fixed FSD approach is still applied in certain situations.

The main advantages of isocentric treatment follow from the fact that the patient need be set up once only rather than for each beam. The advantages are:

1. Stable patient position,

2. Reduced treatment time,

3. More reliable field matching,

4. The option for rotational therapy.

There are some advantages to fixed FSD treatment related to the patient being further from the treatment head. However, in most cases these are not significant compared to the greater stability of the isocentric set up.

1. There is more flexibility in positioning the treatment head allowing a greater range of beam entry positions.

2. Lower scatter dose from the treatment head.

3. Higher relative depth dose within the beam reduces entry dose.

4. Larger treatment fields are possible.

The introduction of isocentric treatment was hindered by the difficulties encountered in calculating an isocentric dose distribution before the widespread use of computerized TPSs. With only measured fixed FSD data generally available, combining multiple treatment fields each with a different FSD in order to produce a combined dose distribution was not a practical possibility.

10.3.2 Coplanar versus non-coplanar planning

Another technique limitation which existed in the past was the requirement for coplanar or 2D treatment planning. A coplanar plan is one in which the central axes of all beams lie within the same plane. There were a number of reasons that combined to make this the only practical treatment option at the time.

Before CT scanning provided an accurate 3D patient model, manual methods had to be used to measure the patient's external contours. The manual techniques were time consuming and difficult to perform accurately so a single representative contour was used wherever possible. In addition, planning computers did not have the software or resources to perform 3D calculations quickly.

If we are restricted to a 2D patient model, the only plane in which the geometry can be modelled correctly is the plane containing the beam central axes as this is only plane not crossed by divergent ray lines which enter the patient at adjacent planes. This leads us to the requirement to keep the beam axes coplanar.

Once a full 3D model of the patient is available this restriction is removed allowing greater flexibility in the choice of gantry angle, couch and collimator rotation. Non-coplanar planning has allowed improved dose distributions to be delivered at some treatment sites although in many cases the best solution is coplanar as this will often minimize the path length of the beams through the patient. However, even within a coplanar plan, the 3D patient model is essential in order to calculate the distribution in planes away from the central axes.

10.3.3 Isodose distributions

TPSs use a number of display methods to illustrate dose distributions within a patient but the most usual is the isodose distribution. An isodose line is a line joining points receiving the same absorbed dose to water in a similar way that contours on a map join points of the same height. In a 3D representation, the isodose line becomes an isodose surface.

Isodose distributions from simple beam arrangements incident on a water phantom form characteristic patterns. Familiarity with these simple cases is useful in understanding clinical dose distributions.

The isodose distribution from a single 6MV photon beam incident on a water phantom at 100cm SSD is shown in Fig. 10.1. The distribution is normalized to 100% at d_{max} on the central axis. The isodoses illustrated are at intervals of 10% with the 95% shown in addition.

Features to note are:

a) The increase in spacing between the isodoses on the central axis as the depth increases. This corresponds to the reduction in the gradient of the depth dose curve.

b) The widening of the penumbra with depth.

c) Increased rounding of the isodoses with depth.

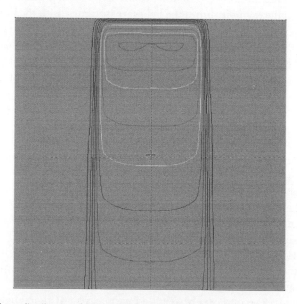

Fig. 10.1 Isodose distribution from a 6MV photon beam incident on water at 100cm FSD (100% is the innermost line). This figure is reproduced in colour in the colour plate section.

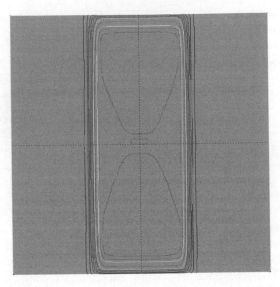

Fig. 10.2 The dose distribution in water resulting from an evenly weighted parallel pair. The beam energy is 6MV and the phantom width 16cm. This figure is reproduced in colour in the colour plate section.

The last two effects act to increase the field size required to cover a target volume by a specified isodose level as the depth of the target within the patient increases.

Fig. 10.2 illustrates the dose distribution produced by two opposed fields — a parallel opposed pair. In the example shown the fields are isocentric with the isocentre placed at mid plane in the phantom. The fields are evenly weighted delivering the same dose at the isocentre where the distribution is normalized to 100%. The isodoses are at 10% intervals with the addition of the 95% and 103%.

The entrance dose from one beam adds to the exit dose from the other so that the distribution obtained will depend upon the energy of the radiation and the width of the patient. If the beam energy is well matched to the width the hot spots will be comparatively low with good dose coverage at mid plane.

In the example shown the phantom width is 16cm and the energy is 6MV. The hot spot is less than 105% and 95% dose coverage is maintained at mid plane.

A fairly common variation on the standard parallel pair involves weighting one of the beams higher than the other in order to shift the high dose region towards one surface. The example shown in Fig. 10.3 is a 2:1 weighted parallel pair, the 2:1 weighting referring to the relative dose contribution from each beam at the mid plane. The distribution is normalized to 100% at the isocentre. The isodoses are at 10% intervals with the addition of the 95%, 105%, 110% and 115%. The dose distribution in this case is deliberately uneven. There is some sparing on the lower dose side but at the expense of a significantly larger hot spot.

The addition of a second evenly weighted parallel pair at right angles to the first results in what is often referred to as a four field brick. The high dose area becomes a square or rectangle defined by the region where the four fields overlap. An example for the case of equally sized fields is shown in Fig. 10.4.

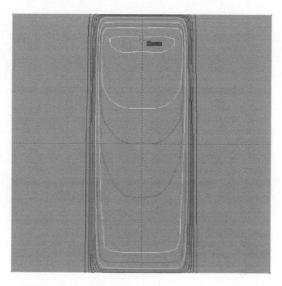

Fig. 10.3 Isodose distribution resulting from a 2:1 weighted parallel pair. This figure is reproduced in colour in the colour plate section.

The advantage is a large reduction in entrance dose compared with the parallel pair. In this example the entrance dose is the same for all fields because the path length to the isocentre for each beam is equal. Entrance dose will increase as the path length to the isocentre increases but can be reduced by using a higher beam energy.

It is often preferable to use three fields rather than four in clinical situations. An example would be pelvic radiotherapy with an anterior and two lateral fields.

Fig. 10.4 Isodose distribution resulting from a four field brick field arrangement. This figure is reproduced in colour in the colour plate section.

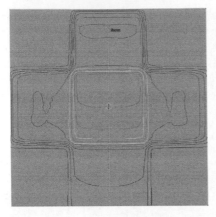 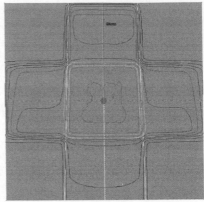

Fig. 10.5 Isodose distribution resulting from three plane fields at right angles (left) and for the same set up with 45 degree wedges added to lateral fields (right). This figure is reproduced in colour in the colour plate section.

The example on the left in Fig. 10.5 shows the isodose distribution produced on a square phantom by three plane fields at right angles. The anterior field is weighted higher than the laterals to equalize the entrance doses. The posterior region of the phantom receives only exit dose from the anterior field. The result is a steep dose gradient within the high dose region perpendicular to the lateral fields, due to the attenuation of the anterior field. The dose gradient can be removed by using wedge filters (see Chapter 8) on the lateral beams.

Wedge filters tilt the isodose curves within a field so that they no longer lie perpendicular to the beam's central axis. The angle through which the isodose curves are tilted is known as the wedge angle. On older treatment units, the wedge filter was a wedge shaped metal absorber inserted into the beam. Four wedge angles were typically available: 15, 30, 45 and 60 degrees. On modern linear accelerators (linacs), a greater range of wedge angles is created using either a single motorized wedge in combination with an open field or by moving a collimator through the treatment field during delivery.

The distribution on the right in Fig. 10.5 shows the result of adding wedges to the plan on the left. A 45 degree wedge is added to both lateral fields with the thick end anterior. The wedges remove the dose gradient but tend to increase the entrance dose at the thin end.

10.3.4 Tissue compensators

Dose gradients perpendicular to the central axis will also be produced if a beam is incident on the patient obliquely or when the surface contour is curved. In this case the gradient is due to differences in the path length through the tissue. A wedge filter can be used to compensate for the different path lengths.

Wedges can provide only very simple compensation in one dimension. In general, changes in patient contour will occur in both directions perpendicular to the beam and with varying gradients. Isodose adjustment in this case requires a custom made two dimensional compensator.

To design such a compensator, the amount of missing tissue must be assessed along each ray line from the source to the compensation plane and material introduced into the beam to provide the extra attenuation required. Compensators may be made from any material but aluminium is often used. If the compensator is placed too near the patient, skin sparing will be lost due to electron contamination and so it is advantageous to attach the compensator to the treatment head. Compensators are becoming obsolete due to the increasing availability of IMRT (Section 10.4.2).

10.3.5 Inhomogeneous media

So far, consideration has only been given to the dose distribution within a homogeneous water phantom. In a patient, tissue densities vary and this will affect the distribution of dose. Tissues with a lower electron density than water attenuate the beam less allowing greater transmission of radiation and the reverse is true of materials with a higher electron density.

The relative electron density of lung lies in the range 0.2 to 0.3 while that for bone is typically 1.2. For MV radiation, these numbers relate to how beam attenuation is modified from the case of transmission through water and lead to the idea of equivalent path length. In terms of beam attenuation 1cm of water is equivalent to approximately $(1/0.25) = 4$cm of lung or $(1/1.2) = 0.8$cm of bone.

10.3.6 Volume definition

To treat a patient with cancer with radiotherapy, at a minimum we need to be able to identify which regions of the body we want to direct the radiation at (targets) and which parts to avoid. There are many ways this can be done, however, it makes sense if there is an international consensus on naming and volume definitions. This has many advantages, allowing comparison of treatments between different departments and countries.

Box 10.1 Who are the ICRU?

Conceived in 1925, this group of international volunteers' primary objective was to propose an internationally agreed upon unit for the measurement of radiation as applied to medicine. They now publish guidance on a wide range of radiation issues including volumes.

The International Commission on Radiation Units (ICRU) published a report (ICRU 50) standardizing the prescribing, recording and reporting of photon beam therapy in 1993. This was supplemented by additional definitions in a subsequent report (ICRU 62) in 1999.

The volumes defined can be spit into the following categories:

- Malignant disease,
  - Gross Tumour Volume (GTV),
  - Clinical Target Volume (CTV),
  - Planning Target Volume (PTV),

- ◆ Treated Volume,
- ◆ Organs at Risk,
- ◆ Irradiated Volume.

First let's consider the malignant disease definitions, these can be thought of as concentric circles in 2D or spheres in reality, increasing in volume (Fig. 10.6).

10.3.6.1 GTV

The GTV is the 'gross demonstrable extent and location of the malignant growth' i.e. the tumour. The GTV may also be extended to encompass metastatic lymph spread. So, we need to make sure that the GTV is always in the treatment beam!

However, there may in some cases be no GTV: e.g. if it has been removed by surgery.

10.3.6.2 CTV

The CTV includes all of the GTV (if present, or at least where it was prior to surgery) and any subclinical malignant spread, i.e. all of the cancer. This is the volume that must be fully treated, i.e. should always be in the treated dose region, if one is to achieve the aim of radical radiotherapy.

Both the GTV and CTV are volumes that one could mark on the patient, if the disease was only skin deep. As the patient moves, the CTV moves because it is all part of the patient. This is an important point, because the CTV moves with the patient, we consider it from the patient's point of view. It is called being in the patients 'frame of reference'. However, the CTV can also move with the organ it is in, e.g. lung cancer. Hold on to this thought, we'll come on to moving CTVs soon!

10.3.6.3 PTV

Now, the GTV and CTV are patient volumes, but we are aiming to treat them with an external beam treatment unit. For this example we'll use a linac. Unfortunately linacs are not perfect machines, from day to day there will be small differences in how they operate due to mechanical imperfections. Additionally, as the patient lies on the couch each day, even with very good immobilization, there will be small differences in where they are lying (we are looking for millimetre precision . . .).

The patient will be treated on many days and we want to make sure that the CTV is in the high dose region, each time the patient is treated. As such, we need to take account of all the variables that can affect the accuracy of the treatment delivery. Put simply, we combine all these together: patient set up variation, likely CTV movement, linac mechanical variations etc and deduce a safety margin to grow the CTV by. By growing the CTV by this amount we create the PTV. The aim is that by irradiating the PTV we will always be irradiating the CTV. It is worth noting that if the CTV is close to the skin surface the PTV may actually extend outside of the patient.

The size and shape of the PTV dictates how the linac will be set to treat the CTV. As such, it is linked to the linac not to the patient. The PTV is a region in space around the linac isocentre. The technical (ICRU) speak for this is that the CTV and GTV are in the 'patient frame of reference' and the PTV is in the 'linac frame of reference'.

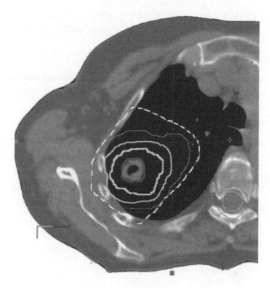

Fig. 10.6 Lung tumour showing from the inside line, GTV, CTV, ITV, PTV, treated volume (dotted line). OARs are spinal cord and PRV. Shown more clearly in colour in the colour plate section.

The PTV is an often misunderstood volume, it is not linked to the patient as such. If you consider the CTV (for a moment) as a basketball, then the PTV is analogous to the hoop!

As we alluded to earlier, we need to irradiate a larger area than the CTV (PTV) to be sure of always treating the CTV. As you may have noticed, the day to day variations that affect the CTV could be split into two main groups: those that are due to the patient's internal anatomy and those external to the patient. This subtlety is addressed in ICRU 62, with two additional volumes defined to help classify the contributions to the margin added to the CTV to create the PTV. These are the internal target volume (ITV) and the set up margin (SM).

10.3.6.4 ITV

The CTV shape can be influenced by adjacent organs, e.g. the bladder and rectum filling both impact on prostate location and shape. This change in CTV size/shape is assigned a label in ICRU language as the internal margin (IM). If we add the IM to the CTV we get the internal treatment volume (ITV). The ITV is meant to encompass the CTV as it varies with the patient's internal anatomy.

10.3.6.5 SM

The set up margin encompasses all the other factors that could lead to the CTV being missed by the treatment field. These are factors that are external to the patient. As such, they are considered relative to the linac. This can be reduced by good patient immobilization and/or use of online correction.

These two additional volumes give us the following sequence:

Patient centred volumes: GTV -> CTV -> ITV

Treatment centred volumes: ITV+SM -> PTV

There is nothing complex about these two extra volumes, they are merely added to help clarify where the data comes from to justify the growth of CTV to PTV.

10.3.6.6 Treated volume

In most cases this is the volume of the 95% isodose. In an ideal world this would be the same as the PTV, but almost always it is bigger.

10.3.6.7 Organs at risk (OAR)

All tissue not in the CTV is considered to be 'normal tissue' and the dose to this should be kept as low as possible. However, not all normal tissue is equally sensitive. Organs that are known to be especially radiation sensitive are usually outlined in the TPS as OAR. Normal tissue tolerance levels need to be considered when prescribing radiotherapy, based on normal tissue complication probabilities (NTCP) and clinical evidence.

10.3.6.8 PRV

The OAR move with the patient and are susceptible to the same random and systematic errors as the CTV. As such, we grow the OAR volume by a set amount to create the planning organ at risk volume (PRV). Consider this a safety margin around the sensitive structure.

10.3.6.9 PTV meets PRV

There may be times where the PTV and PRV overlap. At this point it is up to the clinician to decide which volume takes priority.

10.3.6.10 Irradiated volume

This is simply quoted as the volume of tissue that receives a dose that is considered significant in relation to normal tissue tolerance.

10.3.6.11 A note on bolus

There are occasions (e.g. in the treatment of head and neck tumours) when the CTV extends up to the skin surface. The build up effect inherent in high energy photon beams may lead to under dosing of part of the CTV (inappropriate skin sparing!). To remove this effect, a thin sheet of water equivalent material, usually 5 or 10mm thick, may be placed on the skin: this is known as bolus. The aim is to ensure that any beam build up effects occur in the bolus, not in the CTV.

Note The presence of some essential pieces of equipment (e.g. the treatment couch and immobilization shells) are likely to provide a bolus effect when a radiation beam passes through them. This effect can be pronounced and the build up effect (and associated skin sparing benefits) can be unintentionally eliminated.

10.3.7 Dose prescription

The purpose of dose prescription is to turn relative dose (in percentages) shown on the treatment plan into absolute dose in Gray (Gy). It also enables calculation of treatment monitor units.

A single point in the relative (percentage) distribution is chosen (usually on the 100% isodose) and this is assigned to the prescribed dose. Now all the percentage isodoses can be expressed in Gy. This point is called the ICRU reference point.

The ICRU recommend that the PTV be covered by the 95% isodose, with the maximum less than 107%. This is commonly interpreted as 95 to 105% covering the PTV.

10.3.7.1 ICRU reference point

The system most commonly used for reporting dose is that based on ICRU50/62. A reference point is chosen within the PTV and is referred to as the ICRU reference point.

The ICRU define this point as:

1. The dose at the point should be clinically relevant,
2. The point should be easy to define in a clear and unambiguous way,
3. The point should be selected so that the dose can be accurately determined,
4. The point should be in a region where there is no steep dose gradient.

Ideally, the ICRU reference point should be on the isocentre which should be in the centre of the PTV. However, this is not always possible. As a minimum, the reference point should be in the PTV and conform to the four criteria above.

10.3.7.2 Dose reporting

Tumour control depends on the dose to the CTV. However, the dose to the CTV can only be estimated with regards to the dose to the PTV. Hence, the doses to the PTV are commonly reported. However, if the PTV gets close to, or goes outside, the body contour, a modified PTV should be produced for reporting purposes. This reporting PTV will give a more realistic minimum dose to PTV value.

As a minimum, one should report the dose to the:

♦ ICRU reference point,

♦ Maximum Dose to the PTV (ideally this should also be the maximum patient dose),

♦ Minimum Dose to the PTV.

The ICRU define three levels of dose reporting, of which the above is level 1.

Level 2 assumes that computerized planning is in use and that GTV, CTV, OR, PTV and PRV can be defined and the doses to them calculated relatively accurately.

Level 3 is for developmental techniques.

It is possible to produce a lot more information than this about a treatment plan. Indeed, for advanced treatments, such as IMRT, one would desire more information than this to decide if a plan was suitable for patient treatment! As such, dose volume histograms (DVH) are produced (explained later). These, in conjunction with the 3D dose distribution help to decided what is an acceptable plan and form the record of the patients treatment.

10.3.8 Simple monitor unit calculations

10.3.8.1 What is a monitor unit?

A monitor unit calculation allows you to work out how much radiation the linac should produce to deliver a certain amount of dose to a specific position within a patient. These three concepts are central to radiotherapy:

1. The monitor unit—this is a measure of the amount of radiation passing out of the head of the linac.
2. Prescribed dose—the amount of radiation dose (Gray) to be delivered to a point.
3. The part of the patient that should receive this dose—where this point is located dictates how many MU per Gray are required.

Note Photons transfer energy to the patient, but it's electrons that deliver the dose! So dose delivered to tissue is dependent on the energy a photon has (to liberate an electron) and how tightly the electron is bound to the atom (type of material you are measuring dose in).

So this means:

$$more\ photons = more\ scattered\ electrons = more\ dose$$

A monitor unit can result in a large range of doses depending on where one specifies the measuring point. As an extreme example, if you are in the control area of the linac then 1 MU results in virtually no dose to you. However, if you were to be looking right into the head of the linac when I MU was delivered you would receive quite a large dose! Hopefully, after this chapter you should be able to calculate this dose yourself. Now, here's an equation for basic MU calculation (DON'T PANIC).

$$MU = Dose\ per\ fraction\ (Gy)/Output\ Dose\ (Gy\ per\ MU)$$

The linac generates photons from a very small source, we'll consider it to be a point with no dimensions. From this point, deep in the head of the linac, the photons produced spread out . . .

Most don't make it, stopped by the lead shielding and tungsten jaws. However, the ones that do fan out as they leave the head of the linac and speed into the treatment room.

Dose is defined as the energy absorbed by a fixed amount of material (Joules per kilogram). So when measuring dose we need to keep our measuring volume the same size. To continue with the previous example, if you are looking into the linac head we can consider the dose to the lens of your eye.

This brings us to the first thing that changes the relationship between MU and dose: if our measuring point (or eye!) is very close to the radiation source, all the photons will be close together, therefore a lot of photons will pass through our measuring volume (lens), giving a lot of dose. However, if we move our measuring volume far away from the linac head (e.g. stand against the far wall) then the photons will have spread out considerably by the time they get to us, meaning relatively fewer photons will pass through our measuring volume, giving less dose. This is called the inverse square law.

So,

1   Absorbed dose is reduced as you move away from the radiation source.

2   The other main factor affecting the number of photons getting to your prescription point is how much matter (water, tissue, bone etc.) the photons have to pass through to get there. The photon beam is attenuated (photons get knocked out) as it passes through matter.

This is the reason is safe to stand in the linac control room, as there is sufficient material between you and the radiation to block it from getting to you.

If you put points 1 and 2 together they describe what is happening to produce a percentage depth dose curve (PDD).

10.3.8.2 Monitor unit calculations

In order to calculate monitor units, you first need to know what the reference conditions are for the department (or exam!) you are working in.

Assuming calibration conditions are:

100cm SSD, Reference depth= d_{max}, 10×10cm field, source to reference measurement distance = 100 + d_{max}

Therefore, 1cGy = 1MU at a source to measurement point distance of 100cm+ d_{max} and an attenuation depth of d_{max}. This is the situation that links MU to dose and all MU calculations are taken from this point. All we are doing when calculating MU is adding various factors to move us from this point (where we know that 1cGy requires 1MU of radiation) to the clinical situation which we wish to know the required MUs to deliver our prescribed dose. If anything in your clinical situation is different from the calibration conditions, then you need a factor to correct for it.

10.3.8.3 General MU equations for single field treatments

Isocentric (100cm from source to prescription point)

MU = 100 × (Dose/fraction)/TMR × FSF × TrayFactor × WF × $((100+d_{max})/100)^2$

Fixed SSD (100cm from source to patient surface)

MU = 100 × (Dose/fraction)/PDD × FSF × TrayFactor × WF

10.3.8.4 Why multiply by 100?

Why is the first term in the MU equation 100? Remember 1MU give 1cGy under calibration conditions. However, radiation dose is prescribed (in the UK) in Gy. 1Gy = 100cGy, hence the 100× multiplier.

10.3.8.5 Which equation to use?

The main difference between the two equations is one uses PDD and the other TMR. Which one to use depends on the treatment set up.

10.3.8.6.1 When to use PDD

Fixed SSD (usually 100cm) treatments lock the distance from the source to the skin. So the source to reference point distance (SRPD) must be variable. Therefore, the measuring point moves. This is exactly what a PDD is! So we use PDD tables. These include attenuation and inverse square reduction in dose.

10.3.8.6.2 When to use TMR (Tissue Maximum Ratio)

Isocentric treatments have a fixed SRPD of 100cm, therefore variable SSD. So for each MU calculation we need to get from reference conditions (d_{max}) to the depth required (100-SSD) without changing the SRPD: this is what a TMR is! So one uses TMR data as this mimics the treatment, includes attenuation of radiation beam but with no inverse square effects.

PDDs were used for linacs that could not rotate fully around the patient, so every beam was set up at 100cm to the skin surface. TMRs were invented for rotational therapy (isocentric treatments), where the patient remains static and the treatment machine moves, giving a variable SSD with gantry angle. These days, most treatments are isocentric. Only single fields and extended SSD beams use PDDs directly.

10.3.9 Field size factor (FSF)

Also known as 'output factor', this factor increases with field size and its range depends on beam energy and linac design.

Radiation dose to a point in the patient can be thought of as coming from two sources: primary radiation and scattered radiation. Primary radiation is that which comes straight from the source (imagine a straight line connecting your measuring point to the radiation source). This is (theoretically) the radiation from a zero size radiation field (or very nearly zero). The 'scatter' is radiation that has to be deflected (or scattered) from its original path to end up at your measuring point.

At zero field size we have only primary radiation. As the field size increases the primary radiation does not change. However, there is more radiation passing by your measuring point which might end up being deflected. The larger the radiation field is, the more likelihood that scattered radiation will reach your measuring point. Hence, radiation dose increases with field size.

> For linacs producing MV photons, a typical range of field size factors is 0.9 to 1.1, this range is larger for lower energies.

10.3.10 Equivalent square

Clinical treatment requires radiation field shapes that can be square, rectangular, round or (more likely) very irregularly shaped. So, to calculate the MU required for these fields we need to know the PDD or TMR for this particular field shape. However, when measuring beam data on a treatment machine it is not practical to measure all the PDDs/TMRs for every conceivable field shape. All that is measured are these data for a range of square fields (which is much easier!).

So, how do we get the PDD/TMR we require for our MU calculation of a rectangular field? As field size changes, the scatter changes and it turns out that for every irregular field, there is an 'equivalent' square field with the same depth dose characteristics!

Rectangular fields are simplest. The following formula (where A and B are length and width, or X and Y) gives the equivalent square:

$$2 \times A \times B / (A+B)$$

Note The measured data will be for field sizes with nice round numbers (4cm, 6cm, 8cm etc.). However, the equivalent square you calculated will rarely be for a nice round number. To get the PDD or TMR you need, you will have to interpolate between the nearest two numbers to your equivalent square.

10.3.10.1 Equivalent square: extensive shielding

In the case where your radiation field is more complex than a rectangle then more complex methods must be employed to deduce the equivalent square. The most versatile (and least likely to be employed in a manual calculation) is a 'sector integration' (also known as a Clarkson Integral). In this method the radiation field is split up into segments with your measuring point at the centre. Divided up like a cake, each wedge of field (or cake) will be a different size, but all will be an equal angle of cut (Fig. 10.7). If you now consider each segment to be part of a circular field of that radius, then you know how much scatter that would give to the centre. If you do this for all segments, then divide the amount of scatter by the number of segments, you will come to an answer for the equivalent square.

10.3.10.2 Equivalent square/circle

The mean radius gives the radius of the equivalent circular field. To convert to an equivalent square you multiply the radius by 1.77.

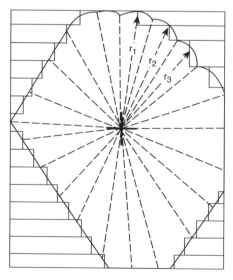

Fig. 10.7 Example of scatter integration calculation, with the field split into segments of equal angle. The radii of the first 3 segments (r1, r2 and r3) are shown.

10.3.11 **Monitor unit calculations: worked examples**

Box 10.2 Case 1

Applied single field, 6MV, 8cm x 20cm field size, 100cm SSD, to deliver 8Gy in 1 Fraction to a point at 5cm deep on the central axis

    SSD=100cm, therefore use Fixed SSD equation. Only one field, therefore field weight = 1

    MU = 100 × (Dose/Fraction) x field weight/(PDD x FSF x TrayFactor x WF)

    Equivalent Square = 2×A×B/(A+B)

    = 2 × 20 × 8/(20 + 8)

    =11.4

Now use equivalent square to look up PDD

Table 10.1 6MV PDD

Depth (cm)		Equivalent square field size			
		9	**10**	**12**	**15**
	1	98.8	99.0	99.1	99.4
	2	98.2	98.8	98.0	98.3
	3	93.9	94.6	94.1	94.2
	4	89.8	90.1	90.4	90.2
	5	85.3	85.8	86.3	86.3
	6	81.3	81.7	82.0	82.6
	7	77.2	77.7	78.1	78.7
	8	73.0	73.9	74.4	75.1
	9	69.3	70.0	7.06	71.5
	10	65.8	66.4	67.1	68.2

    Look along the horizontal row for depth of 5cm. Then interpolate between field sizes 10 and 12 to get the value for 11.4cm equivalent square: PDD = 86.2%. We need to change this into a fraction to use in the calculation (i.e. how it varies from a value of 1) so 86.2% (of 1) is 0.862.

    Look up the FSF using the X and Y jaw values: Gives FSF = 1.01

    No lead blocks in beam, so no lead tray, therefore: TrayFactor = 1.00

    No Wedge in field, therefore: WF=1.00

    So MU required to deliver 8Gy in 1 Fraction to 5cm deep with a 100cm SSD field

    MU = 100 × (8/1) × 1/(0.862 × 1.01 × 1.00 × 1.00)

    = 800/0.871

    = 918 MU

Box 10.3 Case 2

Isocentric single field, 95cm SSD, 6MV, 9cm × 20cm field size, to deliver 8Gy in 1 Fraction to a point at 5cm deep on the central axis.

SSD=95cm, depth=5cm, SSD+depth = 100cm, therefore use isocentric equation

$MU = 100 \times (Dose/Fraction) \times$ field weight/TMR × FSF × TrayFactor × WF × $((100+d_{max})/100)^2$

Equivalent Square = 2×A×B/(A+B)

= 2 × 20 × 8/(20 + 8)

=11.4

Now use Equivalent square to look up TMR

Table 10.2 6MV TMR

		Equivalent square field size		
		10	11	12
Depth (cm)	1.4	1.000	1.000	1.000
	2	0.997	0.997	0.997
	3	0.972	0.973	0.974
	4	0.946	0.947	0.949
	5	0.917	0.918	0.921
	6	0.889	0.891	0.894
	7	0.861	0.864	0.868
	8	0.833	0.834	0.839
	9	0.803	0.805	0.809
	10	0.771	0.775	0.781

Look along the horizontal row for depth of 5cm. Then interpolate between field sizes 11 and 12 to get the value for 11.4cm equivalent square: TMR = 0.919

Look up the FSF using the X and Y jaw values: Gives FSF = 1.01

No lead blocks in beam, so no lead tray, therefore: TrayFactor = 1.00

No Wedge in field, therefore: WF=1.00

So MU required to deliver 8Gy in 1 Fraction to 5cm deep with a 95cm SSD isocentric field:

$MU = 100 \times (8/1) \times 1/(0.919 \times 1.01 \times 1.00 \times 1.00 \times ((100+1.4)/100)^2)$

= 800/(0.928 × 1.028)

= 838 MU

Note It has taken 8% fewer MUs to deliver the same dose to the same point when treating with an isocentric set up. This is because the patient is closer to the treatment machine in the isocentric set up. Therefore the radiation is more concentrated (inverse square law) at the reference point — so more dose per MU.

10.3.12 **Field weight**

For MU calculations with more than one field, we need to take into account the amount of dose delivered by each field when calculating the MUs. In the case of a parallel opposed treatment to deliver 4 Gy to the mid point of the patient, we would usually choose to have half the radiation delivered from each treatment field. Therefore, when calculating the MUs we need to workout how to deliver $4 \times \frac{1}{2}$ from each field (the $\frac{1}{2}$ is the field weight). Similarly, if we had 4 treatment beams 90° apart, all with equal contribution to the total dose (of 4Gy), then each beam would only have to contribute $\frac{1}{4}$ of the total dose. Therefore, we would calculate the MU required for each beam by multiplying by $\frac{1}{4}$ (field weight $= \frac{1}{4}$).

If the field weighting is unequal, this is easily dealt with. For example, for a four field plan if one wishes to give more radiation through the anterior and posterior fields than through the lateral fields, one could have 0.3 weight on anterior and posterior but only 0.2 weight on both laterals. The important thing to note is that all weights must add up to 1.

10.3.13 **General MU equations for multiple field treatments**

Therefore, the MU equations for multiple fields become:

Isocentric (100cm from source to prescription point)

$$MU = 100 \times (Dose/Fraction) \times field\ weight/TMR \times FSF \times TrayFactor \times WF \times ((100+d_{max})/100)^2$$

Fixed SSD (100cm from source to patient surface)

$$MU = 100 \times (Dose/Fraction) \times field\ weight/PDD \times FSF \times TrayFactor \times WF$$

Note It is very unusual to have multiple field fixed SSD treatments if one has isocentric treatment machines, as treatment is much faster when only one patient set up is required.

Box 10.4 Case 3

Isocentric parallel opposed fields.

There are two, equally weighted fields. That is, we want each field to give the same radiation dose. Therefore, we will calculate the required MU to deliver half the dose from each field (this is why the beam weight is 0.5).

Calculate the MU required from each field to deliver 2Gy to the centre of the patient.

Patient separation = 15cm; Field size = 15cm × 10cm (X jaw, Y jaw); Energy = 6MV

Isocentric (100cm from source to prescription point)

$MU = 100 \times (Dose/Fraction) \times field\ weight/TMR \times FSF \times TrayFactor \times WF \times ((100+d_{max})/100)^2$

Equivalent square = $2 \times 15 \times 10/(15 + 10) = 12.0$ cm

Depth = mid point of patient

= separation/2

= 15/2

=7.5cm

Look up TMR in 6MV TMR table for depth of 7.5cm and equivalent square of 12cm

Box 10.4 (continued)

Table 10.3 6MV Field Size Factors (FSF)

		X(cm)				
		12	15	18	20	25
	8	0.992	0.997	1.000	1.001	1.004
Y(cm)	9	1.000	1.005	1.008	1.009	1.013
	10	1.010	1.012	1.013	1.015	1.019
	12	1.015	1.021	1.024	1.026	1.031

TMR = 0.854 (0.8535 rounded to 3 decimal places)
Y jaw = 10cm, X jay = 15cm, therefore:
FSF = 1.012
No lead, no wedge, So: TrayFactor = 1.00 and WF = 1.00
So MU required to deliver 2Gy to depth of 7.5cm with 2 opposed treatment fields:
MU (per field) = $(100 \times (2/1) \times 0.5)/(0.854 \times 1.012 \times 1.00 \times 1.00 \times (101.4/100)^2))$
= 100/0.8886
= 112.5
= 113 MU per field (rounded to nearest monitor unit)

10.4 Principles of CT treatment planning

CT imaging has been the staple of treatment planning for many years and for good reason. With it we get accurate spatial information and electron density information. Therefore, when modelling the patient treatment, we know how far the radiation has to travel and what it has to travel through.

10.4.1 Image manipulation and image fusion

Patients may be imaged with more than one modality MRI/CT/PET each with its own advantages (and disadvantages). TPSs allow one to merge these data sets so that they overlay each other. This process goes by the title of fusion or registration. With this complete, one can draw around structures that are clear on one data set and have them appear on another data set where they would not be so easy to outline (e.g. MRI brainstem volume transferred to CT).

10.4.2 Defining the volume, growing tools

Now you have a detailed 3D representation of the patient on the computer screen in front of you, with (hopefully) enough contrast between tissues to allow you to identify the GTV and CTV. TPSs have various drawing tools available to allow you to mark the relevant treatment volumes on the virtual patient anatomy. To generate the PTV one needs to be able to grow the CTV. This can be done automatically on most TPSs, either growing the CTV uniformly in all directions, or more in one direction than others. Care must be taken at this stage as any errors may impact on the clinical outcome: leading to missing the tumour or over irradiating normal tissue. A similar volume growing process is carried out with the OAR, some of which are grown to make PRVs.

10.4.3 Beam's eye view

Planning systems offer a feature called a beam's eye view (BEV). This allows you to view the patient as if from the treatment machine head. You can switch patient structures on or off (e.g. body contour, PTV, OAR) as you require. This is very useful for shaping the MLC and jaws to ensure coverage of the PTV and also avoidance of the PRV (Fig. 10.8).

10.4.4 Margins

How large must a treatment field be to give adequate dose coverage to the PTV? A common mistake made by novice planners is to fit the MLC or jaws up to the edge of the PTV in the BEV. However, the beam edge is defined by the 50% isodose, so your coverage at the edge of the PTV will be too cold! A dose profile across a 6MV field shows that you achieve 95% of the maximum dose about 0.6cm inside the field. So, best to leave a 0.6 or 0.7cm margin between your PTV and field edge (Fig. 10.8).

10.5 Advanced concepts, developments and special techniques

10.5.1 Plan verification and evaluation

10.5.1.1 Isodose display

When the calculation is complete, the TPS computer will link together regions of equal dose, called isodoses. These are analogous to contour lines on a map. When a TPS performs a calculation, it is actually calculating dose to a 3D series of points called

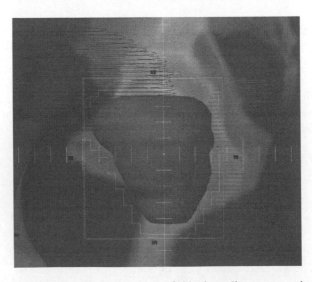

Fig. 10.8 BEV showing PTV, bladder and MLC field edges. Shown more clearly in colour in the colour plate section.

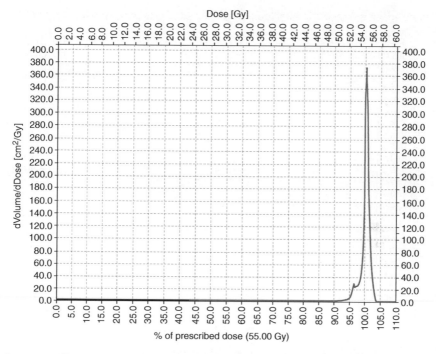

Fig. 10.9 Differential DVH showing PTV coverage.

a dose grid which covers the patient anatomy treated. Isodoses are interpolated from dose grid points but the doses to individual grid points can be used to generate Dose Volume Histograms (DVH), a valuable plan assessment and comparison tool.

10.5.1.2 Dose volume histograms

A DVH is a 2-dimensional graphical representation of the 3-dimensional dose distribution for individual organs. It is useful for evaluating and comparing treatment plans. However, DVHs do not replace the full isodose distribution as they do not contain geometric information: they can not tell you where in the organ the dose is.

There are two types of DVH in use in radiotherapy.

- Differential (frequency) DVH,
  - Fractional volume (v_i) of organ receiving a dose (D_i)
- Integral (cumulative) DVH
  - Fractional volume (v_i) of organ receiving a dose (D_i) *or greater*

The Differential DVH shows the homogeneity of dose to a structure (Fig. 10.9). In the case of a PTV one could look at the width of the peak (narrow is good) and the maximum and minimum doses are easy to see. However, this DVH is not so useful for OAR.

The Integral DVH is the one most commonly used to give PTV and OAR data (Fig. 10.10). The two most common uses of this DVH are to see the:

1. Global maximum received by an organ (serial organs), such as spinal cord maximum.
2. Dose received by a certain volume of an organ (parallel organs), such as V20 of lung.

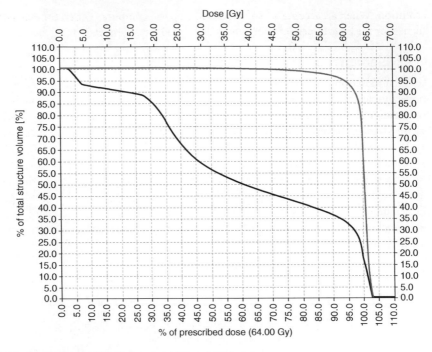

Fig. 10.10 Integral DVH showing PTV and OAR data.

10.5.1.3 Digitally reconstructed radiographs (DRR)

The CT data used for the treatment plan can be used for yet another purpose: generating images to mimic plain X-ray radiographs. For each treatment field in the plan, one has the option to generate a DRR. For this, the TPS takes the treatment field geometry and draws rays through the CT set along the beam path. The CT electron density information allows the TPS to create areas of low and high contrast, such as a planar X-ray beam would do on a film/detector. This DRR can then be used to compare against X-ray images taken on the treatment unit to verify the accuracy of the patient set up. DRRs are used to replace simulator images in virtual simulation. Fig. 10.8 has a DRR with PTV, OAR and field shape superimposed on it.

10.5.2 Elements of inverse planning

3D conformal treatments are generally forward planned. That is, the planner decides on all the treatment parameters (field size, beam weight, gantry angle etc.) then calculates the plan to see what the dose distribution looks like. If they don't like the result, they can change something (e.g. beam weight if peripheral dose is too high on one side) and try again. This is an iterative process by the planner. This works well when we are dealing with simple beam modulation options: field size, field weight and wedges. Inverse planning is similar to this, but the computer program makes the iterative changes and decides on whether it likes the result based on criteria the planner tells it. These criteria can be in the form of DVH graphs. The planner tells the algorithm what the maximum and minimum acceptable doses are and the computer attempts to find a solution. When an acceptable solution is found, the computer generates maps of fluence for the linac to deliver using the MLC to modulate the dose.

10.5.3 Elements of intensity modulated radiotherapy

An open radiation field can be said to be of uniform intensity. If we want to create more complex distributions than these we have to be able to change the radiation dose to the patient coming through different parts of each field. The simplest way to do this is with a wedge. However, to create curved isodose distributions (to bend round spinal cord) we need to be able to modulate the radiation field in 2 dimensions. This is achieved by moving MLC leafs whilst the beam is on. Each opposing pair of MLC leaves work together to create a strip of modulated dose. If the leaves are closed, then nearly no dose gets through (so we can create an area of low dose) as the leaves are opened more radiation gets through. So the size of the gap between the leaves dictates how much radiation gets to each part of the treatment field (Fig. 10.11). If we have

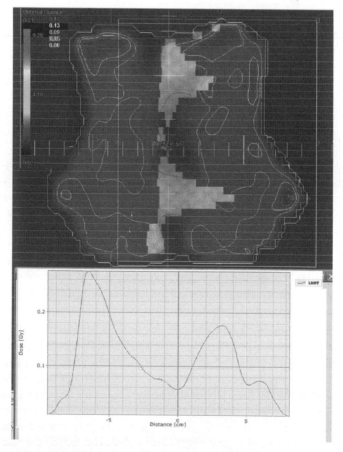

Fig. 10.11 BEV of MLC half way through an IMRT field. Dose profile taken through vertical axis. This figure is reproduced in colour in the colour plate section.

60 pairs of leaves moving across the radiation field together, we can create very complex radiation fields. A typical IMRT plan uses between 5 and 9 radiation fields at static gantry angles, equally spaced (See also Radiotherapy In Practice – External Beam Therapy, Chapter 3).

10.5.4 VMAT

Volumetric Modulated Arc Therapy (VMAT) uses the same concepts as discussed for IMRT but uses two extra variables to allow us to delivery complex dose distributions. While the beam is on, the gantry rotates, we can vary:

♦ Gantry rotation speed,

♦ Dose rate (certain vendors only),

♦ MLC position.

In this way we can deliver dose distributions similar to IMRT plans but in a significantly shorter treatment time.

10.5.5 Tomotherapy

Tomotherapy units are based on the design of CT scanners, they could be said to look like a large CT scanner. They have a short 6MV waveguide spinning around the patient (much as the KV source in a CT scanner does) which produces a fan beam of 6MV photons. This beam is modulated by a mini MLC. The dose is built up in a helix as the patient moves through the gantry on the couch. The tomotherapy unit can deliver complex dose distributions to large volumes in this manner.

10.5.6 Stereotactic radiotherapy

Stereotactic radiotherapy is not very different from all the rest of radiotherapy, in that the process is largely the same and the radiation is inherently the same. However, this treatment is generally used to treat very small volumes to very high doses in a small number of fractions. This technique aims for tumour ablation with minimum dose to surrounding tissue therefore we require highly collimated radiation beams (very small fields with sharp dose fall off at the edges) and very precise patient positioning (sub millimetre localization). There are many different treatment machines (and linac add ons) available for stereotactic treatments but all are aimed at achieving these two objectives. Stereotactic treatments were originally used for brain lesions, but now the modality has branched out to treat other parts of the body, known as stereotactic body radiotherapy (SBRT), such as lung and liver. For this we must consider organ motion and manage with gating or appropriate ITVs. Treatment can be delivered on a gamma knife, linac (with or without micro MLC), Cyberknife or proton beams.

10.6 QC in treatment planning

10.6.1 Commissioning the TPS

A commercial TPS will come with calculation algorithms already installed but not configured. If you have worked in more than one radiotherapy centre you will notice that every department does things slightly differently. As such the TPS must be customized to the local environment. It is up to the local user (physicist) to supply the computer with the local information (how is an MU defined in your hospital?) and test the system to ensure it works safely. This process can take several months to fully implement and test, involving (among other things) the entry of dosimetric data measured on the treatment machines.

The testing regime of a TPS is an involved procedure which is beyond the scope of this book. But physicists endeavour to check every conceivable variation of treatment parameters to ensure there are no hidden horrors to discover in the future! Just because a TPS is FDA approved or CE marked does not automatically make it safe!

Every time the TPS software is upgraded (annually on most systems) the physicists go through a mini version of this commissioning process.

10.6.2 QC of CT scanner on TPS

The CT scanner provides us with data crucial to accurate dose calculation. Such as the physical dimensions of the patient and the electron density of their tissues. This allows the TPS to calculate the dose from a radiation beam within a patient taking into account the patients' bone, lung etc. Therefore, if we are to trust this information we should regularly test the CT scanner to ensure it measures distance and electron density correctly. A simple way to do this is to scan a plastic block (phantom) with small inserts of known electron density, separated by a known distance. One can then import these CT scans onto the TPS and test that the known values are reproduced on the TPS.

10.6.3 Checking of individual patient plans

10.6.3.1 Conformal

A physics check of a conformal plan should at a minimum check that the patient details are all correct; the moves from tattoos to treatment centre are correct; the plan is a 'good plan' (i.e. dose distribution covers PTV and avoids PRV); plan is for the correct dose and fractionation, all local processes have been followed and (finally) that the MUs are correct. The MU check should be a fully independent calculation of the MU required to deliver the prescribed dose to the prescription point in the patient.

10.6.3.2 IMRT/VMAT

The checks above are all necessary for these more complex plans, however there is an extra test we need to run due to an extra risk factor that these treatments have. IMRT plans require the MLC to move during treatment and what you see on the TPS is an interpretation of how the TPS thinks the MLC will work. The resultant dose distribution in the patient is totally dependent not only on the MLC working correctly, but also on the TPS algorithm modelling the unique leaf motion for this patient correctly. As such, all patient IMRT plans are verified on a linac.

10.6.3.2.1 *Measuring IMRT plans —*

There are various devices available to do this task, but the fundamental requirement is that you have an independent dosimetry system that can measure the dose delivered by an IMRT plan on a linac and then software to read this device and compare the dose distribution to what you expected (i.e. what the TPS tells you to expect). There is some debate about whether this is necessary, but at this point in time it is a small price to pay to avoid a very large error if things go wrong.

The devices used for IMRT dose measurement (also called verification) can be split into 2 categories: 2D or 3D.

The 2D devices can have continuous detectors: either film or the linac's portal imager, or be part of a discontinuous array: a grid of ionization chambers or a grid of diodes.

The 3D devices currently available are not truly 3D (more 2.5D), but involve using a large number of diodes arranged such that there will always be an accurate measurement taken from a radiation beam no matter what the gantry angle is (this is useful/essential for VMAT verification).

10.6.4 Calculation algorithms

10.6.4.1 Manual planning

Even up to the early 1980s, many radiotherapy treatment plans were produced by hand. For most departments, the use of CT for treatment planning was a far off dream. Target volumes were localized from plain AP and lateral radiographs and transferred to a single patient external contour that had been acquired usually by using thick wire. Measured or calculated single isodose beams were selected to cover the target and combined together by copying onto tracing paper and adding together manually. As can be imagined this was a little tedious.

10.6.4.2 The first planning systems

In the early 1980s, TPSs started to appear. These computers essentially did the same job as hand planning but a lot quicker. The type of dose calculation method used was known as a 'beam library model' — these systems took account of contour correction but ignored missing tissue. So for breast and head and neck type treatments, the systems tended to overestimate doses actually received by tissue. Density corrections for lung etc. were possible but crude and without CT, it was difficult to know where the lungs were anyway.

10.6.4.3 Simple planning systems

In the late 1980s, a new type of dose model appeared called 'scatter integration', using sector integration as discussed in Section 10.3.10 above. Unlike the beam library model, these differentiate between primary radiation (from the treatment head) and scattered radiation (generated within the patient). This model was able to take account of missing tissue generally improved the accuracy of treatment planning. Such scatter integration models are still used today, mainly in programs used to independently check the accuracy of planning system dose calculations.

Scientists continued to work to develop dose calculation methods that more closely represented the true nature of dose deposition in patients. Remember: photons interact and electrons give dose.

10.6.4.4 What are we trying to model?

All current treatment planning algorithms are an approximation of what is actually going on in the patient. From what you learned about photon and electron interactions you already know more than you think about how the whole process takes place.

Essentially, high energy photons (primary radiation beam) interact with matter leaving a slightly lower energy photon (scatter) and a moving electron (scatter) broken free from its atom. The photon goes on to liberate more electrons etc. All these electrons move through the material/tissue losing energy (scatter) to other electrons (scatter) causing 'dose' to be deposited. Everything we are discussing here is trying to model this process. Specifically, we are trying to follow the electrons (because they deposit dose). So, how do we model photon interactions with electrons — and what happens to them after the interaction? The best way is to use a technique called 'Monte Carlo' modelling. This is based on the random nature of interactions: there is a probability an interaction will occur, not a certainty. So every photon (and there are millions and millions of these in a simple treatment field) must be tracked through every interaction and the products of every interaction (photons and electrons) must be tracked too. This is a huge pyramid scheme where the number of particles being tracked rises enormously and, before you know it, your computer has died of boredom (or something more technical).

So whilst we have computers unable to do this in a reasonable time we must find short cuts to simplify (speed up) this calculation. However, it should be noted that despite lengthy calculation times, Monte Carlo calculations are considered the 'gold standard' in terms of dose calculation and are usually used as a benchmark for new dose models that appear. It should be noted that with the ever increasing development of computer processing power, real time Monte Carlo calculations are a probable rather than a possible development in the future.

10.6.4.5 Pencil beam algorithms

In the mid 1990s 'convolution' or pencil beam convolution (PBC) type models became available. These start off along the lines of Monte Carlo but use much quicker methods to get to a result of dose deposited in a patient. The 'pencil beam' model is based on kernels . . . These are created in advance, usually during the commissioning of a planning system. They are calculated once (using Monte Carlo modelling) and then used for each patient calculation. Therefore the very time consuming calculations have been taken out of the patient treatment path.

10.6.4.6 Kernels

Kernels are simply a map of the dose that will occur for your energy of radiation beam (e.g. 6MV) hitting a small volume of water. A kernel looks bit like an asymmetric ellipse, elongated in the direction of travel of the radiation beam (Fig. 10.12).

Then the scatter from any patient geometry can be recreated by adding up these pre-calculated small scattering volumes into the shape you require—a bit like using Lego.

So, one dose kernel by itself is not much use and we need to add many of them together to produce a dose distribution for a single beam. A typical radiotherapy plan will consist of a large number of small volumes of tissue, each of which will release energy and deposit dose in surrounding small volumes of tissue. The process of adding all the kernels together is known mathematically as convolution, hence the name convolution dose model.

Fig. 10.12 Photons hitting small box of water give resultant scatter shown as ellipses. Intensity of scatter falls with distance from box.

10.6.4.7 Pencil beam convolution: pitfalls

This is all very well in a homogeneous (i.e. water) medium. However, what happens when you introduce a change in density? The classic example is radiation passing through lung (Fig 10.13). The PBC will scale the kernels (scatter) along the path of the beam, but not laterally. So, in a breast treatment the reduced scatter from low density lung is not modelled so the PBC overestimates the dose to the breast tissue next

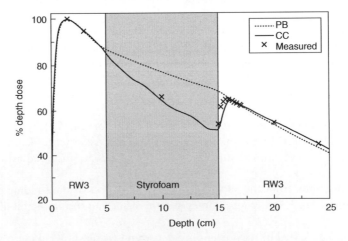

Fig. 10.13 Graph showing how %DD through lung (styrofoam) are calculated by pencil beam (PB) algorithm and superposition algorithm (CC) versus measurement. Reproduced from A. Nisbet, I. Beange, H. Vollmar, C. Irvine, A. Morgan & D. Thwaites (2004). Dosimetric verification of a commercial collapsed cone algorithm in simulated clinical situations. *Radiotherapy & Oncology*, **73**, 1: 79–88, with permission of British Institute of Radiology.

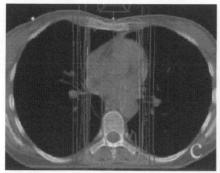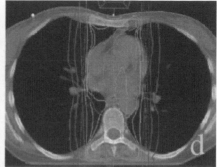

Fig. 10.14 The same patient treatment calculated with pencil beam, c, and superposition, d. The more accurate modelling of scatter in d can be seen by the low dose levels (darker grey lines) spreading laterally, but the high dose lines (lighter grey lines) being drawn in medially. This figure is reproduced in colour in the colour plate section. Reproduced from C. Irvine, A. Morgan, A. Crellin, A. Nisbet & I. Beange (2004). Clinical Implications of the Collapsed Cone Planning Algorithm, *Clinical Oncology*, **16**, 2: 148–154, with permission from The Royal College of Radiologists.

to lung. In this sense, they only model the heterogeneity along a line and are known as 1D models.

10.6.4.8 Superposition algorithms

The next level of algorithm uses convolution and superposition of pencil beams. Variants of this algorithm go by such names as 'collapsed cone' and 'AAA' (or triple A as it's spoken) depending on the manufacturer. The main difference is that these algorithms can correct for inhomogeneity in all directions. They are truly 3D. They model the loss of scatter in lung and the build up of scatter (and dose) upon reentering tissue after lung and the broadening of the beam penumbra in lung (Figs 10.13 and 10.14). It does this by scaling the dose kernels, roughly in proportion to the densities of the materials they're passing through. The superposition type models are generally accepted to be the most accurate type of dose calculation model currently available and can accurately model missing tissue and the effects of lung tissue on radiation beams.

Chapter 11

Beam therapy equipment

S J Colligan and J Mills

11.1 Introduction

Clinical megavoltage radiotherapy beams range in energy from 4 to 50MV for photons and 4 to 20MeV for electrons. These beams provide the vast majority of radiation treatment. In addition to electrons, particle beams consisting of protons and ions (Chapter 4) are in use clinically and the energies of these beams are in the order of 100's of MeV. In contrast, kilovoltage beams used for superficial tumours are typically X-rays with energies between 50 and 300kV (see Table 8.1). Historically, X-ray production was initially in the kilovoltage energy range. Efforts to produce high energy X-ray beams started in the 1930's. Large accelerators were built using high voltage potentials to provide a more penetrating radiation beam to deliver a dose deeper into the patient. Eventually in the 1950's the production of high energy photon beams was revolutionized with the use of radio waves and the modern medical linear accelerator (linac) was born. Nowadays the array of technology is extremely impressive and a wide range of ionizing radiation beams can by produced electrically.

11.2 Fundamentals of electrically generated beams

There are two fundamental forces which make electrically generated beams possible: the electrical force and the magnetic force. Using these forces for the acceleration and control of charged particles defines the construction and operation of these machines.

11.2.1 The electrical force

The electric force requires two ingredients: a charged particle (an electron) and an electric field (kilovoltage or megavoltage). The electric potential exists between two plates marked anode and cathode (Figure 11.1), with the anode having a positive electrical potential with respect to the cathode. This sets up an electric field strength (which is dependent upon the voltage and distance) between the anode and cathode.

11.2.2 The magnetic force

The magnetic force also requires two ingredients, one of which is again a charged particle. However for there to be a magnetic force the charged particle must be moving with some speed, v. The second ingredient is a magnetic field strength, B. The force which acts on the charged particle is applied at right angles to both the motion of the particle and the magnetic field strength.

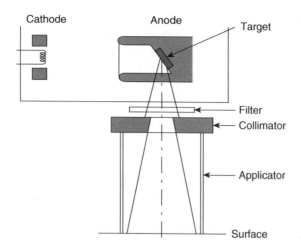

Fig. 11.1 Layout of the major components of a kilovoltage (kV) therapy X-ray machine.

The size of the force is related to both the speed of the particle, its charge and the strength of the magnetic field.

Electric fields are created by voltage power supplies which can be up to thousands of volts. These voltages are then applied across metal plates to create electric fields which cause a charged particle to move (e.g. from the cathode to the anode).

Magnetic fields are created by current power supplies which can be up to hundreds of amperes. These currents are driven through metal coils to create magnetic fields to cause the charged particle to change direction (e.g. bending the electron beam round a corner).

11.3 **X-ray production**

Now we have electrons moving (electrical force) and can control where they go (magnetic force) we can use them to produce an electron beam. The production of an X-ray beam is dependent upon the bremsstrahlung interaction process, which is discussed earlier in both Chapters 3 and 4. To produce an X-ray beam, a target of appropriate material is bombarded by an energetic beam of electrons. As the electrons interact with the target a portion of their energy goes into the X-ray beam but the majority is lost in collisional processes which cause heating of the target. The efficiency of X-ray production never exceeds 10% because great deal of energy is lost as heat. Therefore cooling of the target is essential.

11.3.1 **Target material**

The choice of target material is influenced by the fact that bremsstrahlung production is proportional to the atomic number Z (the number of protons in the nucleus of the material), therefore the higher this number the better. However, the material needs to be in a manageable form, preferably a solid that is robust and machinable in standard engineering processes. Tungsten (a metal with a high Z of 74 and a high melting point)

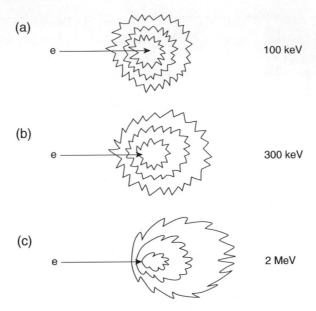

Fig. 11.2 A representative indication of the spatial and directional release of X-rays from an electron beam of increasing energy incident onto a target (e = electrons).

and tungsten alloys are ideal as they can withstand the extreme heating that occurs during the X-ray production.

Knowledge of the directionality of the bremsstrahlung X-ray production process is very important for the design of the target. At superficial energies the bremsstrahlung is virtually omni-directional (Fig. 11.2a). As the energy increases to the order of several 100's of keV there is greater directionality to the X-ray beam (Fig. 11.2b). Ultimately as the energy enters the megavoltage range the X-ray beam is extremely directed forward (Fig. 11.2c). At the low energy, sufficient X-ray intensity can be obtained with the target turned at an angle to the direction of the electron beam. However at the MeV energy level the greatest intensity comes *though* the target and so it is called a 'transmission target'.

Bremsstrahlung production results in X-rays with a spectrum of energies. The spectrum is modified by absorption and attenuation of the beam within the target. However the basic spectrum can be thought of as a continuum up to the maximum energy of the incident electrons (see Chapter 3, Fig. 3.1). There are two other points to consider: firstly the typical effect of filtration which reduces the intensity at lower energies and makes a 'harder', more penetrating beam; secondly, at low energies, in the kV range there are characteristic X-rays which arise from electron transitions in the atom and not from bremsstrahlung (see Chapter 3). These are not detrimental to the clinical use of low energy beams and can be reduced by additional filtration.

11.4 **Charged particle production**

The electrons required to create X-rays and electron beam are generated by thermal emission from appropriate materials.

11.4.1 How are electrons generated for X-ray production?

By heating a metal so that its electrons become very energetic and leave the surface of the metal, an 'electron cloud' can be created. Tungsten, which provided an ideal X-ray target material, can be used for this. Tungsten can be raised to a temperature high enough to form a sufficient electron cloud. X-ray machines of all energies use tungsten *filaments* as the source of electrons for the accelerated electron beam.

11.5 Kilovoltage X-ray beam machines

Orthovoltage and superficial units provide X-ray beams in the kV energy range. The units typically take the form of a tube assembly that can be easily moved to direct the beam onto the patient. The source to surface distance is typically in the range of 20 to 50cm and the field is defined at the surface by a mechanical applicator. An example of such a machine is shown in Fig. 11.1.

11.5.1 Tubestand

The tube is mounted on a 'tube stand' that is attached to the ceiling or the floor. This enables easy movement of the tube to direct the X-ray beam appropriately. Brakes are fitted to the rotational axes and translational runners so the position can be fixed prior to treatment.

11.5.2 High voltage circuits

The intensity of the X-ray beam produced at a particular kilovoltage depends upon the number of electrons emitted from the filament. This is controlled electronically to respond to changes in the mains supply and stabilize the beam current.

11.5.3 Collimation and beam profile

The anode construction provides the initial collimation of the beam as the X-rays come off the target almost omni-directionally (Section 11.3). This is referred to as a hooded anode (Fig.11.3) and the aperture defines the maximum size of conical X-ray beam. The collimation is completed by an applicator which is fixed directly below the tube aperture and provides the following:

- The base of the applicator consists of a collimator that defines the shape and size of the radiation field at the end of the applicator.
- The end of the applicator is the shape and size of the radiation beam.
- The end of the applicator determines a fixed distance from the source.
- The axis of the applicator is mechanically aligned to the axis of the radiation beam.

Most applicators have transparent ends in order that the treatment machine operator can easily view the area to be irradiated and ensure the correct set up on the patient.

The variation of intensity across a kilovoltage beam, referred to as the beam profile is not uniform as there is a significant drop off from the central axis of the beam towards the edge (Fig. 11.4). This variation is never taken into account when planning a patient treatment. However it is important to remember that there is not uniform intensity across the beam and that adequate margin is allowed for effective treatment.

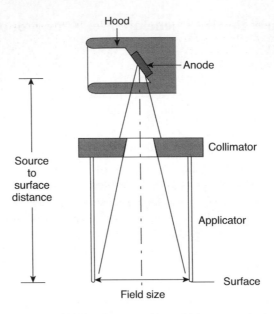

Fig. 11.3 The components of a kilovoltage machine which accurately determine the field size and the source to surface distance.

The beam profile is also affected along the direction of the target angle by the so-called the 'heel effect'. This arises due to absorption of some of the X-rays within the target itself so that there is a slight asymmetry along that axis (Fig. 11.4).

11.5.4 **Beam energy**

The X-ray beam from any bremsstrahlung target consists of a range of energies up to the maximum accelerating potential (kVp) produced by the machine (Section 11.3). Hence for a 120kV machine, 120keV will be the accelerated electron energy and in turn the maximum X-ray photon energy will be 120keV (therefore the beam can be called 120kVp). This spectrum of energies is modified using a filter to remove lower

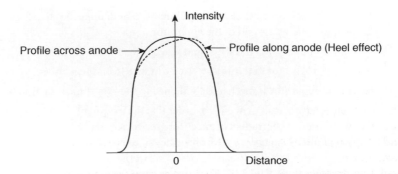

Fig. 11.4 Beam profiles from a kilovoltage machine beam along and across the target.

energy X-rays and so reduce the dose to the skin. Additional filtration is chosen to provide a beam with the desired depth dose penetration. The energy characteristics of the beam are often referred to as the beam 'quality' and depend on both the accelerating potential and the filtration. The beam quality is specified in terms of the half value layer (HVL) which is the thickness of the layer of metal required to reduce the intensity of the beam by half (see also Chapters 3 and 8). The depth dose curve for a kilovoltage beam has a maximum at the surface followed by a decreasing dose with depth and a typical curve is shown in Fig. 11.5 along with the depth doses of two electron beams for comparison. It can be seen that for the electron beams a high dose is maintained across a superficial depth while with the kilovoltage beam the dose rapidly falls away across the same depth. Also, beyond that superficial depth the electron beam dose decreases rapidly, whereas a substantial dose is delivered beyond that depth by the kilovoltage beam. Kilovoltage treatments have been used for many years and clinical practice has built up a great deal of experience with such treatments.

11.5.5 **Control of output**

Either a timer or an ionization chamber (also known as a monitor chamber) is used to control the dose delivered. The timer is normally a count down device and starts when the treatment is initiated. After the set time has elapsed the voltage and current is switched off and treatment terminates.

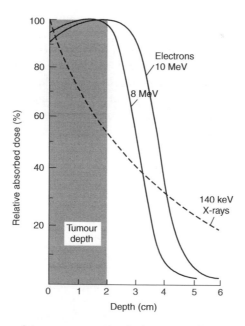

Fig. 11.5 Comparison of the percentage depth dose curves from kV X-rays and electron beams.
Taken from Klevenhagen S. C. (1985). Physics of electron beam therapy, Fig. 3.1.
Bristol: © Adam Hilger Ltd.

Most modern units use a monitor chamber (through which the beam passes), much like a linac. It terminates the beam when the required number of monitor units (defined in Chapter 10) has been reached.

11.5.6 Customized shielding

Beam collimation on orthovoltage machines is achieved by a range of applicators which have regular field sizes and these are chosen by the user to suit the clinical requirements for the unit. To treat lesions of non-standard size and irregular shape lead cut-outs are made to define the treatment field exactly to the area required. The cut-out is used with one of the regular sized applicators to irradiate the treatment area while the surrounding normal skin is shielded. For lesions on the face and close to the eyes, a lead mask can be made based upon a plaster cast of the patient. When the treatment area impinges onto the eye, eye shields can be inserted to provide some protection to the lens. Commercial lead and tungsten shields are available. The problem with eye shields is the contribution from scatter which reaches the region under the shield from the surrounding field. The contribution from this scatter can be significant.

11.5.7 Calibration of dose output

The absolute dose is measured in accordance with a protocol or code of practice. This ensures uniformity of practice between institutions. It uses the absolute dose calibration of the ionization chamber being used for the measurement and this is traceable to a national standards laboratory (Chapter 7). The calibration can be done in air with the use of mass energy absorption coefficient ratios to determine the dose to water or tissue. Alternatively the measurement can be done directly at depth in water. For superficial kV units it is typical to determine the surface dose rate as this is where the dose is to be applied. However for orthovoltage X-ray units it can be preferable to quote the dose deeper than the surface and closer to the target. In the latter case it is preferable to calibrate at depth in water, usually 2cm.

11.6 Cobalt-60 radioactive source machines

The penetrative limitations and high skin dose of kilovoltage X-rays drove the desire for higher energy X-rays. The radioisotope cobalt-60 (^{60}Co) was discovered at the University of California by Seaborg and Livingwood in the 1930's and in Canada in 1951 it was used for teletherapy.

Nuclear reactors provided an ideal production platform for the substance. ^{60}Co produces high energy gamma emission in the radioactive decay process—about four times that of the highest deep X-ray unit photon which is 300kVp. It also has a relatively long half-life of just over 5 years.

The depth dose characteristics of a 10x10cm ^{60}Co beam at 80SSD is compared in Fig. 11.6 to that for a 10cm diameter 300kVp kilovoltage beam. The depth dose characteristics from a linac for 10x10cm field size 6MV and 25MV megavoltage beams at 100SSD are also shown in Fig. 11.6.

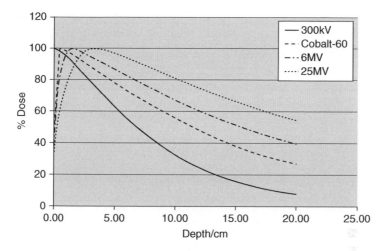

Fig. 11.6 Comparative depth dose distributions along the central axis of X-ray beams as the energy increases.

The forward scattering of the electrons produced from the ^{60}Co photon interactions produces a skin sparing effect with a maximum dose at 0.5cm depth rather than at the surface as with the kV beam, providing an enormous technical advantage for teletherapy.

The cobalt unit represents an extremely simple method to generate a megavoltage treatment beam. There is no need for the highly skilled maintenance and costs associated with linacs and so teletherapy can be provided at low cost where technical support is not readily available. This is the enormous potential for these machines — simplicity and economy.

11.6.1 Radioisotope source

The radioactive ^{60}Co source consists of a small container which encapsulates pellets or discs of ^{60}Co with an activity of typically 200TBq. A typical source is a 2cm diameter by 2cm long cylinder. One of the deficiencies of the cobalt treatment unit is this large source produces a large penumbra in its radiation fields (Fig. 11.7). The source is housed in a shielded safe which has a shutter mechanism to transport the source into an exposed position. Periodic testing is required to ensure that there is no leakage of radioactive material from the source encapsulation. Emergency procedures must be in place to ensure that in the event of a failure of the transport mechanism the patient can be safely removed from the room and the source returned to the safe by manual means.

11.6.2 Beam collimation

The collimation of the ^{60}Co uses a system of primary collimator, secondary collimator and penumbra trimmers as shown in Fig. 11.8. The penumbra of a radiation field depends on three aspects; the size of the source, the distance of the secondary collimator jaw from the source and electron scatter at the edge of the beam (Fig. 11.7). Scatter

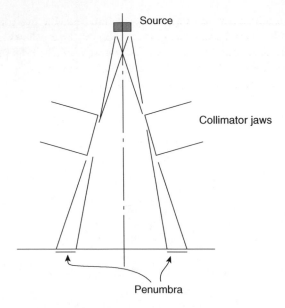

Fig. 11.7 Production of a geometrical penumbra due to the finite size of a radiation source.

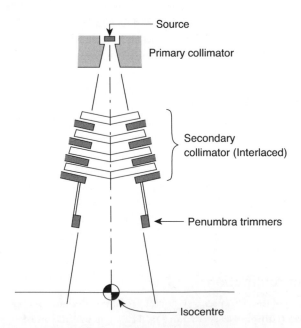

Fig. 11.8 The interleaved secondary collimators and penumbra trimmers used to minimize the size of the penumbra in a cobalt beam

only becomes significant at very high energy photon beams, above 15MV. However a 2cm ^{60}Co source is very large. If then the secondary collimator jaws were placed one above the other the penumbra across the inner jaws would be larger than across the outer jaws. So in order to provide a more equitable size in both directions the jaws of the secondary collimators are interlaced as shown in Fig. 11.8.

11.7 Electromagnetic waves, waveguides and cavities

The modern medical linacs of today have been made possible by electromagnetic (EM) radiation (Chapter 2). It was found that EM radiation could be propagated along metal conduits and these are called *waveguides*. The dimensions of the waveguide can be chosen to control the wavelength and propagation speed of the EM wave. Although the speed of the wave cannot exceed the speed of light, it can be made to travel slower than light and this is of significant value in the design of one type of medical accelerator waveguide. The last essential aspect for acceleration is that the waveguide aligns the electrical component of the wave to the direction along which acceleration is required. As described in Section 11.2, a charged particle will experience acceleration in an electric field.

11.8 Megavoltage linear accelerators

The linac is the workhorse of the modern radiotherapy department, developed by pioneers who developed megavoltage machines in the 1950's. The fundamental components remain unchanged although the performance, construction and control systems have advanced through modern engineering, technology, electronics and computers. Today, such medical linacs provide the vast majority of radiotherapy treatments.

11.8.1 General layout and components

The major components of a linac are an electron source, a source of radiofrequency (RF) electromagnetic waves and an accelerating waveguide. These core components can now be found in some custom machines (such as tomotherapy and cyberknife). This section will concentrate on the isocentric gantry mounted machines (linacs). Besides these major core components, a modern medical linac has a gantry assembly to direct the beam into the patient, and a radiation head to enable beam shaping. For X-ray beams the target, which is bombarded by the electron beam to produce the X-rays, is in the radiation head. Steering and stability of the electron beam requires focussing, bending and steering coils which provide magnetic forces. High voltage and high current supplies are also needed along with vacuum pumping systems and cooling systems in order to create a stable beam. Fig. 11.9 illustrates the typical layout of the major components of a linac.

11.8.1.1 The gun

The 'gun' *filament* (Section 11.4) assembly produces electrons by raising tungsten to a high temperature (through electrical heating).

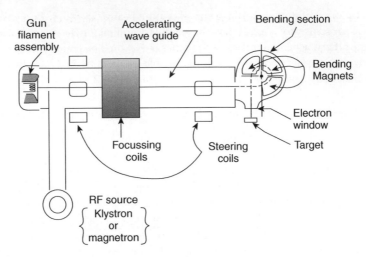

Fig. 11.9 The general layout of the major components found in a linear accelerator producing electron and X-ray beams.

11.8.1.2 The waveguide

There are two types of waveguide: standing wave and travelling wave. Both types use electromagnetic waves at a radiofrequency of approximately 3GHz. In the travelling type the electrons are carried along on a wave (like a surfer at the beach) which accelerates due to the changes in cavity size so the electron and wave move together. In the standing type the wave stays still, but the maximum and minimum points of the wave switch places (like a skipping rope). So as a moving electron passes the maximum of the wave (positive, attractive potential) which was pulling it onwards, this then flips to be a minimum point (negative, repulsive!) which pushes the electron further along. In both cases the electron always sees an accelerating force of a positive (attractive) potential ahead of it and a negative (repulsive) potential behind it.

11.8.1.3 RF wave production

The radio waves are produced by either a klystron or a magnetron. The RF from the magnetron or klystron is transported to the accelerating waveguide through ancillary waveguides.

11.8.1.4 Electrons in the waveguide

The repulsion force between the electrons causes the beam to disperse as it travels along the accelerating waveguide. This is countered by using a magnetic force from focussing coils which are wound around the accelerating waveguide and produce a magnetic field flux parallel to the direction of the electron beam. The focussing coils also steer the beam within the accelerating waveguide.

11.8.1.5 Bending the electron beam

A bending magnet is used to bend the electron beam round to the radiation head. This bending force is also very important in the selection of the beam energy. By altering the electrical current flowing in the bending magnet coil, the strength of the magnetic field can be altered and the appropriate electron energy can be chosen. Steering coils produce a magnetic field to steer the beam into and out of the accelerating waveguide. This maximizes the number of electrons which are accelerated, ensuring the most efficient trajectory of the beam along the guide minimizes the loss from the beam.

The target, all the magnetic coils, the accelerating waveguide, the RF source and large electrical devices such as transformers all heat up during operation. A cooling system is essential to maintain stable and effective beam production. Another aspect of the linac is the need for the accelerating waveguide to be under high vacuum. It is either factory sealed or a vacuum pump maintains the high vacuum required.

The gantry construction is referred to as being 'isocentric'. In effect this means that all the main axes of rotation of the gantry, the radiation head and the patient couch intersect though the same point in space referred to as the isocentre (Fig. 11.10). In practice the isocentre is bigger than a point; it is a sphere of about 1 to 2mm diameter. This is because the intersection of radiation axis and gantry rotation shifts slightly with gantry angle due to the weight of treatment head.

The placement of the patient so the centre of the tumour is coincident with the machine isocentre simplifies treatment with multiple beams.

11.8.2 The X-ray beam

The X-ray beam, originates in a transmission target placed directly in the path of the electron beam immediately after it exits from the accelerating waveguide and the bending section. The exiting X-ray production is collimated into the raw beam by the primary collimator. The raw beam has a peak intensity in the forward direction (Fig. 11.11). Conventionally the aim in radiotherapy has been to deliver as uniform a dose as possible to the tumour. The raw beam is therefore made uniform, or 'flattened'

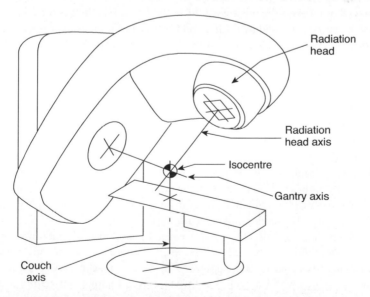

Fig. 11.10 The identified axes of a modern accelerator; gantry, radiation head and couch, showing their intersection at the isocentre.

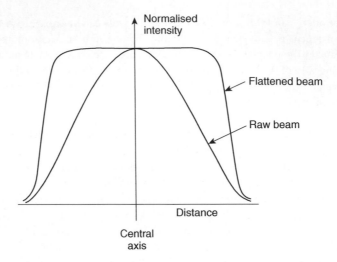

Fig. 11.11 Comparison of a raw MV X-ray beam profile, with the profile following the flattening filter.

using a flattening filter made from steel. The profile of the flattening filter provides the maximum attenuation in the centre and the detailed shape of the filter is specific to the energy of the X-ray beam and the beam profile which it is intended to produce (generally it looks like a witch's hat!).

Following flattening of the beam the X-rays pass through an ionization chamber before going on to field shaping with the secondary collimators. The ionization chamber controls the dose delivered and monitors the beam characteristics. It has three sections and each section covers the full area of the useful beam. The first two sections are the dosimetry channels one and two (Ch-1 and Ch-2). Ch-1 is the primary channel which provides a signal to terminate the beam after a required dose has been delivered (this is monitor units in action!). Ch-2 is identical to channel one and acts as a safety back up should a fault occur with Ch-1. There is also a timer to stop the beam should both ionization chambers fail! The third section of the ionization chamber is a segmented chamber that monitors characteristics of the beam such as uniformity and symmetry. The uniformity refers to the variation of dose across the beam and will be characteristic of the beam energy. The symmetry refers to a tilt in the beam and measures the variation in dose between positions equidistant from the central axis of the beam. The uniformity is dependent upon the energy of the beam and both are dependent upon the alignment of the electron beam onto the target (Fig. 11.12). The uniformity signal can be used to control the gun filament electron emission and the symmetry signal can be used to control the steering coil currents.

Final beam shaping is done by the substantial secondary collimation system. On older machines these were two sets of jaws which could set rectangular field shapes by movement in orthogonal directions. However, all modern machines now have multi-leaf collimation (MLC) with leaf widths at the isocentre ranging from 2.5mm to 10mm. These can easily shape the beam to the target volume and can also vary the intensity of the beam. The intensity can be changed by rapidly changing the beam shape and

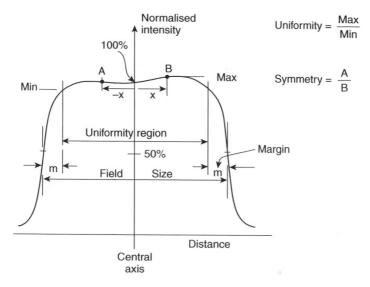

Fig. 11.12 A flattened MV photon beam profile showing the margin and region within which the beam uniformity and symmetry is assessed.

providing multiple beams within the overall field or by dynamic control of the MLC whilst the machine is irradiating (see Chapter 10). Wedges can also be used to modify the uniformity of the beam profile (see Chapter 8).

11.8.3 **The electron beam**

As well as producing X-rays, linacs are also used to deliver electron treatment beams. In this mode, the transmission target is automatically moved out of the beam. The electron beam that is accelerated through the waveguide has a very small size compared with the required treatment field size, so scanning and scattering systems have been developed in order to provide adequate field coverage. Scanning techniques employ magnetic fields which enable the electron beam to be steered across the field. However, the predominant technique uses a dual scattering system, using thin metal foil scatter electrons into a larger area, along with an applicator attachment to the radiation head (Fig. 11.13). In the dual scattering system a primary scatterer is introduced to the electron beam prior to the primary collimator. A secondary scatterer is substituted for the flattening filter and this may be profiled for the specific electron beam. These scatterers and the flattening filter are moved automatically with the energy selection and are specific to the energy selected. The secondary collimators used for the photon fields set an initial field size as the electron beam leaves the radiation head. The electrons are also scattered in air and so an applicator is used to produce a sharp edge to the treatment field at the patient. The applicators have open sides and consist of a set of field trimmers which reduce the field down to the required size. The treatment field is at the normal treatment distance which can either be directly at the end of the applicator if treatment is done with the applicator in contact with the patient or at extended distance from the applicator (usually 5cm). The electron beam passes

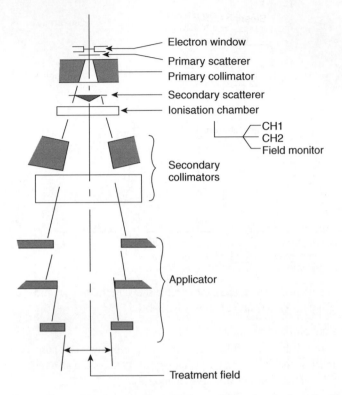

Electron window
Primary scatterer
Primary collimator
Secondary scatterer
Ionisation chamber
CH1
CH2
Field monitor
Secondary collimators
Applicator
Treatment field

Fig. 11.13 The configuration of components in the radiation head when the linear accelerator is used to provide an electron beam.

through the same ionization chamber with multiple sections and segments as the X-ray beam and this controls the delivery of the correct dose and monitoring of the beam as described in Section 11.8.2.

11.8.4 Control systems

Effective and safe performance of a linac depends upon two types of control system. The first controls systems within the accelerator, e.g. the control and tuning of the magnetron, the electron emission from the gun and the steering of the beam. The second relates to the stability controls on voltage and current supplies. A stable current supply to the bending magnet coils ensures that the correct energy has been selected and the correct beam delivered to the patient. Some control systems adjust the beam in response to systems that monitor the beam itself i.e. a dynamic feedback loop. For example, the detection of field symmetry by the ionization chamber in the radiation head. Using the symmetry information the electron beam is steered using the steering coils in order to maintain a symmetric field (Fig. 11.14). An asymmetry in the field is detected by the segments of the ionization chamber in the radiation head. These signals are electronically processed to alter the current supplied to the steering magnet coils and restore the symmetry of the field. Satisfactory beam delivery by a linac relies on the dynamic operation of many control systems. All operate automatically

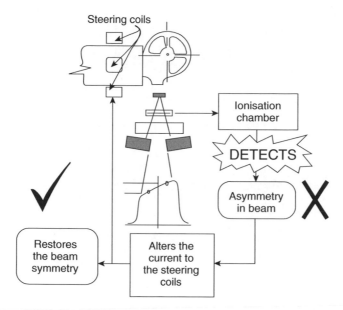

Fig. 11.14 The major components involved in automatically steering the electron beam to maintain a uniform and symmetrical X-ray beam.

and result in interplay between the electron emission from the gun, the tuning of the RF source and the steering of the beam, as well as the stabilization of all the currents and voltages which ensure the correct focussing and bending of the beam.

11.8.5 Alignment of patient to the beam and patient couch

Radiation beams are set up so that they are aligned with the axis of the mechanical rotation of the radiation head (Fig. 11.15). The system for patient beam alignment can be considered to consist of two parts. The first is indicators which are aligned with the axes of the radiation head rotation and hence the radiation beam. The second is indicators which locate the isocentre of the machine in space. In combination the beam alignment to the patient can be achieved and verified.

The indicators of the radiation beam axis are the light field cross wires, illustrated in Fig. 11.15. The light field cross wire projection is achieved through an optical system in the radiation head. A light source is effectively positioned at the X-ray target and cross wires are introduced into the light field using a thin transparent film. With the bulb and the cross wire centre exactly on the radiation head rotation axis an optical projection along the axis of rotation is obtained.

The location of the isocentre is indicated by an optical distance meter (when it shows 100cm) which uses the coincidence between the cross wire projection and the distance meter projection in order to determine distances from the cross ray source to the surface of the patient (Fig. 11.16).

The third indicator is the room laser system which is independent of the machine. The room lasers project three planes of light which intersect as a set of Cartesian coordinate planes with the isocentre at the origin (Fig. 11.16). Surface marks, made during

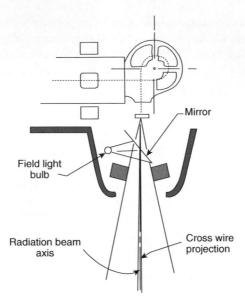

Fig. 11.15 The variation which can exist between the central axis of the radiation beam and the cross wire projection.

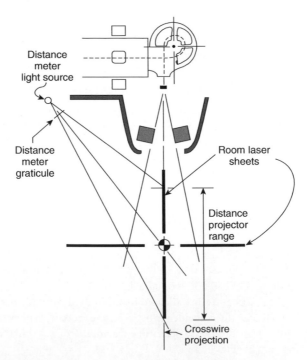

Fig. 11.16 The 'distance meter' system combining projection of graticule and projection of cross wires.

the treatment planning stages of the patient's preparation for treatment, enable the patient to be set up with the isocentre at the chosen location within the patient.

The accurate alignment of the patient couch to the machine is also a crucial part of aligning the patient for treatment. The couch has isocentric rotation in order to allow the target volume within the patient to be oriented to the machine and facilitate the most appropriate alignment of radiation beams for the treatment. The couch then is an integral part of the machine which must perform accurately and reproducibly in order to provide the most effective treatment. Couch indexing has become a routine feature of modern couches. Patient alignment devices lock into the couch in a reproducible way using notches at regular intervals. These locating notches are indexed so that patient set up in these alignment devices can be achieved speedily and with a high degree of accuracy on a fraction to fraction basis within a busy radiotherapy department.

Besides isocentric rotation the couch has three orthogonal movements allowing the patient to be moved vertically, longitudinally and laterally. These movements can be crucial for the matching of treatment field edges in order to produce extended irregular treatment volumes. One such treatment which is extensively undertaken is for irradiation of the breast including the supraclavicular nodes. Virtually all the treatment techniques for this rely on accurate couch rotation coupled with accurate longitudinal and lateral movements in order to match fields without there being any high dose regions due to field overlap.

In practice, the patient is set up to the room lasers as they were at CT planning. Any required moves to get to the planned isocentre (see Chapter 10) are then carried out. The patient should now be in the identical position to their virtual selves on the computer plan. The treatment beams can now be set up and the cross wires and distance indicators should show the same SSDs as indicated on the plan.

11.8.6 Record and verify systems

The machine operators set up the patient and the machine and in some way have this checked and recorded as independently as possible before treatment. Computerized record and verify (R&V) systems were developed to fulfil this role in response to increased treatment complexity and the introduction of Quality Systems that require traceability of processes in radiotherapy. The R&V system consists of three distinct components; storage, check and record.

For each patient treatment, parameters are downloaded from the R&V system to a treatment machine. The machine is then set up by the operators for treatment of the patient and the R&V system monitors the settings on the machine, comparing them to the values stored in the system file. Any differences must be resolved before treatment can start. In exceptional circumstance, parameters can be overridden subject to appropriate checking and supervision. The dose delivered to the patient on each treatment is also recorded (Chapter 5).

11.8.7 Special techniques

There are two major techniques for which the machines are not used isocentrically: Total Body Irradiation (TBI) and Total Skin Electron (TSE) treatments.

For both these types of treatment, the limitation of the accelerator field size is over-come by treating patients at greatly extended distance of up to 4m to achieve full body coverage. For these extended source to patient TSE treatment distances, high dose rates may be specially generated on the accelerator to reduce the treatment time.

Another adaptation of the linac has been to use fixed narrow beam collimators or a micro multi-leaf collimator which can define small fields with dimensions of less than 5cm. These small fields have been used in either a static or arcing treatment techniques for accurate stereotactic radiosurgery and stereotactic radiotherapy (see Chapter 10). An early form of dynamic treatment is achieved by rotation of the gantry during irra-diation, often referred to as arc-therapy. In the case of electrons, this technique can be used to treat a large curved surface such as the side of the chest wall.

11.9 Specialist megavoltage treatment systems

11.9.1 Robotic machines

The isocentric gantry mounting of teletherapy machines is effective in directing multiple beams at the target with relatively simple movements of the gantry and patient couch. Modern robotic control systems coupled with stereotactic radiographic localization of the patient's position allows multiple beams from virtually any direction in space. Such a highly accurate system has been developed with a compact linac mounted upon a robotic arm. The location of the patient is determined from two kilo-voltage images which use bony anatomy or implanted markers in order to track the position of the patient during treatment and correctly align the machine. This flexibility in the multiplicity of beam delivery directions coupled with high accuracy has enabled complex tumour volumes in close proximity to critical organs to be irradiated.

11.9.2 Stereotactic machines

Stereotactic radiotherapy and radiosurgery delivers a high radiation dose to small volumes localized using stereotactic frames in order to ensure a high positional accu-racy of the treatment (see also Chapter 10). A standard linac can be used to deliver such a treatment but the Leksell Gamma Knife is a dedicated machine for this pur-pose. It consists of a treatment cavity for the head around which there is a large array of cobalt sources. The sources are focussed into a small treatment volume, the dimen-sions of which are determined by collimators. The positioning of the patients skull within the treatment cavity is located using a stereotactic frame bolted to the skull to ensure that extremely accurate localization of the compact treatment volume is achieved. For a standard linac machine, a high resolution multi-leaf collimator is used and one commercial system uses stereotactic kilovoltage imaging facilities to ensure highly accurate tracking of the patient and the target.

11.9.3 Dedicated intensity modulated machines

The tomotherapy machine (Tomotherapy Inc.) is completely dedicated to intensity modulation delivery. It has been designed to deliver narrow intensity modulated beams which alter in profile as they are rotated around the patient and in effect com-bine to produce a required dose distribution within that slice of the patient. The patient moves through the machine and adjacent slices are stacked together producing a continuous dose distribution tailored to the clinical requirements for that patient. The tomotherapy machine is based upon the gantry design of a CT scanner and it is also able to generate megavoltage CT images for patient realignment prior to treatment. What is unique about this machine is that it was designed purely for the production of intensity modulated radiotherapy. It is of interest to note that a recent modification of the machine has been to enable it to deliver simple parallel opposed treatments for the breast, perhaps an indication that effective treatment for some regions can be satisfactorily achieved by more simple methods.

The major linac manufacturers have now produced systems which can deliver intensity modulated beams as the machine rotates around the patient. These delivery systems have been shown to considerably speed up treatment delivery compared to the use of multiple static beams. It appears that all IMRT may be performed in this manner in the interest of patient throughput. However while this provides very effective conformal dose distributions efficiently, only tomotherapy is able to simultaneously treat multiple targets over the entire length of the patient.

11.10 Proton and ion accelerators

Proton beam therapy has been undertaken since the mid 1950's, initially with access to experimental beams belonging to physics research institutes. Since 1990 dedicated hospital based treatment machines have been constructed and operated and proton beam therapy is now becoming available throughout the world. Likewise carbon-12 ion therapy which started in the USA in the 1970's on experimental beams has led to dedicated clinical machines since the mid 1990's. As with X-ray production these machines produce the proton and ion beams electrically and for the acceleration forces and the beam guidance they rely on the electrostatic and magnetic forces which were described earlier. Treatment with protons, carbon-12 and other ions can be referred to as Hadron Therapy.

11.10.1 Cyclotron and synchroton

Acceleration is mainly achieved using two types of system: cyclotron and synchrotron.

11.10.1.1 Cyclotron acceleration

The cyclotron was devised by the US physicist Earnest Lawrence who was awarded the Nobel Prize in 1939 for that contribution to physics. Acceleration is provided by the electrostatic attraction between two plates, referred to as the D's, while containment of the beam is provided by a magnetic force produced by a magnetic field which is perpendicular to the plane of the acceleration. The beam experiences an accelerating force between the D's and acceleration occurs up to the point at which the beam is extracted from the cyclotron.

11.10.1.2 Synchrotron acceleration

During the 1940's Edwin McMillan in the US developed the synchrotron in order to overcome limitations of the cyclotron. The synchrotron consists of a circulation path of components parts following an injection section or even an inner acceleration ring (Fig. 11.17). The multiple components provide for acceleration and steering of the

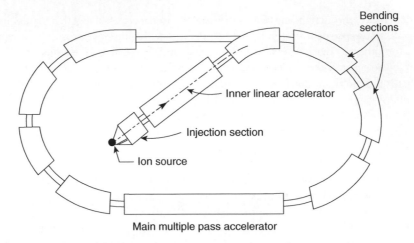

Fig. 11.17 A typical layout for a Synchro-Cyclotron.

beam by repetitive passage of the beam through the acceleration sections as compared to the integral construction of the cyclotron. By variation in magnetic field strength and frequency, even including detection of the beam pulse, it is possible to increase the energy and the flexibility of the particle acceleration.

11.10.2 **Depth dose and spread out Bragg peak**

The clinical attraction to use ions lies in the characteristic depth dose along the axis of beam penetration into tissue. This characteristic Bragg peak distribution (see Fig. 4.6) enables dose to be deposited up to a certain depth but not beyond. The depth of the Bragg peak is dependent upon the energy of the beam with a very crude rule of thumb being 10MeV for every 1cm of depth.

11.10.3 **The treatment head**

The treatment head is often referred to as the nozzle. There are two types which are used to cover the depth of the target volume (see section 4.5). Both types have collimation in order to protect the patient from unnecessary dose outside the treatment beam.

11.10.3.1 Mechanical depth modulation spread beam nozzle

The stages are quite discreet. The first features which the beam encounters is a primary scatterer which broadens the beam and then the rotating depth modulator which spreads the Bragg peak over the range required for the treatment of the target. This is followed by a secondary scatterer which spreads the width of the beam further. The final two features are designed for the specific target. The aperture shapes the beam and the compensator ensures that the deep cut off edge of the beam follows the contour of the target at depth. The target is then irradiated without delivering any dose beyond it. However there are superficial volumes which will receive a treatment dose.

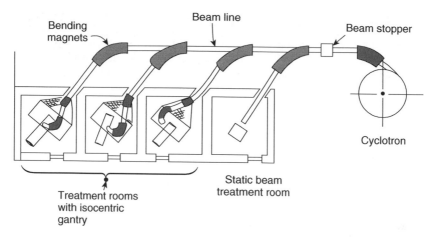

Fig. 11.18 A typical layout for a proton treatment facility with a common cyclotron.

11.10.3.2 Energy depth modulation scanned beam nozzle

The scanned beam uses energy variation to modulate the depth at which the Bragg peak deposits maximum dose. It uses the magnetic force described above to deflect the beam in two orthogonal directions in a plane at right angles to the central axis of the nozzle. This enables, for a given energy, or depth of Bragg peak, the beam to be scanned across a plane or slice of the target at that depth with dose deposited in discrete packets. Layer by layer the entire target can be scanned with the treatment dose delivered effectively to only the target.

11.10.4 Treatment centres

Treatment centres consist of all the standard facilities for patient care and treatment preparation. However, the typical arrangement for the treatment rooms is different, illustrated in Fig. 11.18. A single accelerator generates a beam which is isolated from the main beam line using a beam stopper. Off the main beam line there may be several beam lines to separate treatment rooms. The number of rooms is usually four, but ultimately depends on the treatment time and patient throughput. Most modern facilities have one static beam line treatment room used for treatments in which a horizontal treatment plane is sufficient and the patient can remain upright, and three isocentric gantry treatment rooms.

11.11 The machine lifecycle: procurement to decommission

Treatment machines have lifecycles with specifically identified stages. Clear identification of these stages has provided a structured process to ensure that the performance of the machine is explicitly understood and can be maintained throughout its life. It provides an agreed programme by which both users and manufacturers can deliver machines fit for clinical purpose throughout their life.

11.11.1 Specification and procurement

The first stage is the development of a specification for the machine required and engagement with manufacturers. The specification requires input from all staff groups involved with the machine. It should embody all the legislation and standards applicable for the machine and the specific local requirements from the staff groups. These items can take two forms. Either it asks for an unequivocal requirement, e.g. a source to axis distance of 100cm or it asks the manufacturer to describe a feature, e.g. describe the maintenance support for machine breakdown including the response time for assistance and the timescale for the supply of replacement parts. The responses from the manufacturers should be reviewed by the staff groups and it may be that each group's response can be weighted to give more importance to one groups view. Two aspects are very significant in the choice of a machine: beam matching and transfer of patients. These will tie the procurement choice to the existing equipment if only one machine being sought. Back up and machine transfer are vital to the operation of a busy department and to avoid the interruption of treatment that a different machine manufacturer can be considered only if there are two or more machines being purchased. Architects, builders, computer networking providers and additional equipment suppliers with regard to interfacing should be consulted.

11.11.2 Install, acceptance and commissioning

The next three stages bring the machine through from delivery to full clinical operation.

11.11.3 Installation

During the installation cooperation between the manufacturers installation staff and the local engineers and physicists ensures efficient installation of the machine and that local requirements are met. There can also be beneficial cooperation with the builders during this period.

11.11.4 Acceptance

Following the installation, completion of a formal customer acceptance protocol (CAP) officially marks the handover of the machine from the manufacturer to the purchaser. The CAP documentation and process should be discussed and agreed with the manufacturer before installation. The manufacturer will provide a CAP document and the purchaser may enhance this with additional checks.

11.11.5 Commissioning

Once the purchaser has accepted the machine, an extensive set of measurements are undertaken to fully characterize the performance of the machine for clinical use. These include depth dose, tissue phantom ratios, output factors, tray and wedge factors et cetera and also all the data and verification measurements for dose prediction in treatment planning systems. If the machine has been supplied as being matched to an existing machine, these measurements can be simplified to a comparison with the existing data. So, commissioning the machine concerns gathering all the relevant data from which accurate doses can be delivered during the subsequent treatments which the machine will provide. Guidance with regard to the acceptance and commissioning of linacs is provided in a publication by the Institute of Physics and Engineering in Medicine (IPEM) as well as the British Standard for such machines performance BS EN 60976.

11.11.6 Quality control and planned maintenance

The lifetime of a linac is generally accepted to be 10–15 years. During this time the performance of the machine which was identified at commissioning is maintained through a planned maintenance programme and monitored with a quality control programme. Quality control is also required following repairs to the machine in order to ensure performance has been restored. In the Republic of Ireland, prior to there being a planned maintenance programme, the downtime of machines was about 8% due to breakdown. Following the implementation of a programme, even though the time lost to treatment was 8% including a breakdown of 1.5%, it was possible to schedule the patients to suit the time lost and so the service was improved. Quality control demonstrates that the required performance is being achieved as well as occasionally detecting that the performance has deteriorated below a level at which corrective action should be taken. Planned maintenance and quality control is recognized now in the statutory legislation relating to ionizing radiation, IRR1999 and IRMER2000. It ensures that the computed dose predictions with which an individual patient's treatment has been planned can be actually delivered.

While quality control programmes confirm that machine performance is maintained, some treatment delivery is now too complex to design a programme which could verify all the combinations of treatment. For example the number of field shapes which can be provided by a multi-leaf collimator (MLC) is limitless as is the case for intensity modulated delivery of a component field for IMRT/VMAT (see Chapter 10). Therefore patient specific quality control is required to ensure that every individual patient treatment can be verified. Patient specific quality control is now a component part of quality control of the machine, using fluence measurements to ensure that modulated treatments will be accurately delivered for a specific patient.

Guidance on quality control programmes can be found from the IPEM and for linacs also from the British Standard supplement to BS EN 60976.

11.11.7 Replacement and decommission

Radiation issues in decommissioning is only significant in machines that contain a radioactive source. Manufacturers are obliged to take into account disposal of the machine at the end of its life and the recycling of components where practicable by a European Directive. From a clinical operations point of view, replacing a machine without service interruption is only possible when an additional treatment room is available to enable the new machine to be installed before the original machine is removed. One difficulty with this can be the level of necessary radiation protection. To achieve this all the treatment rooms need to be designed for the highest energy, which could be too expensive.

11.12 **Treatment verification**

Over the past 20 years it has been recognized that simply ensuring the performance of the treatment machine alone is not enough for safe and effective patient treatment. Individual treatment needs to be verified independently of the machine. This is borne out by the several incidents which have occurred over the years. In the past, verification that the correct treatment had been given took place from the clinical indications which were apparent at a patient's treatment review. This is very inadequate, requiring close monitoring, many patients and a long timescale.

It is not sufficient in modern radiotherapy to check that the bullet is satisfactory before it is fired. It is essential to know that the correct bullet was delivered to the correct place.

11.12.1 **Positional verification**

To ensure the radiation beam delivery has been given to the correct location, devices have been developed to verify the location of the patient and even the location of the internal anatomy. Slow radiographic film was used for many years to produce an image of the exit beam from a patient. This has been almost completely replaced by electronic portal imaging devices (EPIDs). These were initially developed to use the megavoltage treatment beam but now systems use a kilovoltage X-ray beam. Kilovoltage beams provide greater contrast between soft tissue and bony anatomy than the megavoltage beam and given that bony anatomy is used to verify the position of the patient, the kilovoltage beam provides more usable images. However bony anatomy is usually a surrogate for the location of the target which is soft tissue. One method to improve this is the use of markers, such as gold seeds, implanted surgically into the target. These enhance the use of planar EPID images for positional verification but they do not definitively identify the soft tissue of the target. This is achieved using a cone beam kilovoltage CT system. This uses a standard kilovoltage imaging source in a position orthogonal to the treatment beam. From images acquired during a rotation of the gantry, a CT data set of the patient anatomy can be reconstructed enabling the soft tissue to be delineated.

For proton and ion machines there is no exit beam. However positional verification is possible using positron annihilation; dosimetry verification can also be achieved through this.

Some other positional verification devices have been developed but these rely on surface markers although one device uses a radio marker which can be implanted surgically into the target. In the case of surface markers, the devices use technology from computer aided design to obtain positional information in three dimensions and track them with time. The relationship between the internal and the external anatomy is a potential pitfall, but mathematical models are being developed to relate these and improve the tracking accuracy. However at present cone beam CT offers a definitive verification of the soft tissue target location.

11.12.2 **Dosimetric verification**

The other aspect of treatment verification is that the correct dose has been delivered. This would ideally be verified by a measurement at the target, however except for superficial targets this is not possible. Instead it is possible to measure the incident dose on the patient from each treatment beam using small solid state dosimetry devices such as thermoluminescent dosimeters (TLD's), diodes and MOSFETS (metal oxide semi-conductor field effect transistor). These measurements are referred to as in-vivo dosimetry (IVD) and the majority of such measurements for routine treatments are performed with diodes. Usually such systems are permanently installed in the treatment room and the operators will place the dosimeter in the incident field.

Obtaining reasonably accurate measurements for dose verification does not come without a cost. Usually the dosimeter will exhibit performance variations associated with the beam such as field size, source to surface distance and the angle of beam incidence. Once the direct reading obtained from the measurement has been corrected for these variations a value of incident dose can be obtained. The calibration of the devices has to be routinely checked and the systems have to be maintained and periodically replaced due to wear and tear.

Nevertheless IVD provides a final verification that the machine has delivered the correct dose to the surface of the patient and thus increases confidence that the patient is receiving the correct treatment.

11.12.3 Transit dosimetry

Transit dosimetry in effect combines both positional and dosimetric verification into a single process. In transit dosimetry the exit megavoltage fluence from the patient is measured by the EPID and used along with the patients CT data to calculate the dose which has been deposited as the beam passed through the patient. By undertaking this process for all beams it is possible to determine a predicted sum total dose in the CT data set of the patient. This prediction of delivered dose distribution can then be compared with the dose distribution that it was intended to deliver.

The EPID device needs to be calibrated in order for it to accurately measure the fluence and relate this to the dose deposited in the patient. However transit dosimetry provides both positional and dosimetric verification in a single measurement and so is perhaps the most complete technique by which treatment verification can be achieved. As discussed above some patient treatments have reached a stage of complexity that it is not possible to rely on the routine quality control of the machine performance to provide adequate confidence in the treatment delivery. For these complex deliveries, only transit dosimetry offers an adequate method of verifying, *in-vivo*, that the correct treatment has been actually given.

11.13 Adaptive radiotherapy systems

Radiotherapy treatment has long been a static process. The target was identified and considered to remain the same for the entire treatment. It was of course known that inter-fractional changes in size and position, and intra-fractional motion would compromise this. Therefore a margin would be allowed which would accommodate anticipated movement of the tumour during the radiation or as the treatment of the disease progressed (see Chapter 10). The development and introduction of EPID devices made assessment of target changes at a chosen fraction a possibility. This enabled tighter margins around the tumour so that normal tissue could be spared and tumour dose escalated. So treatment had moved away from being a static process to being one which could track the dynamics of the target and modify the delivery of treatment to follow that.

Adaptive radiotherapy is the term that covers this dynamic treatment process. It also covers all systems and techniques that track changes in the target and accommodate them in treatment delivery. Adaptive radiotherapy is under development but systems exist that attempt to deal with tumour motion detection and motion compensation. As these developments lead to clinical studies the advantages and the clinical situation for which they are suitable may appear.

11.13.1 **Classical adaptation**

There has always been some degree of adaptation made to treatments. These have mainly related to observed clinical changes which it was intuitively felt could affect the treatment outcome. For example, weight loss in a patient apparent in a change in SSD observed at set up would trigger consideration of whether the target was still being adequately irradiated. Similarly if the tumour could be observed to change such that the target was reduced or increased would require consideration. The problems of replanning a patient have grown as the sophistication of the planning process and the complexity of the treatments have increased. When a single patient outline with orthogonal radiographs was used to plan then the time and resource required for the replan was extremely low by comparison to the need for a new full CT dataset, the outlining of gross tumour volume through to planned treatment volume, shaping of the fields and balancing the required dose limits in the dose volume histograms.

In order to reduce the workload some assessment criteria should be provided by which to filter out the cases for which a replan is definitely required. Such a technique is to consider the changes in SSD observed and whether the net change at the isocentre exceeds the variation expected in normal treatment. Such a simple method is adequate for simple changes. It says nothing about the coverage of the target and is too simple for complex treatments which may for example involve field matching and the possibility of overlap.

11.13.2 **Inter-fractional motion**

Inter-fractional motion is dealt with by standard positional verification systems described in Section 11.12.1. These mainly consist of EPID systems using bony landmarks or implanted markers. There are also stereotactic X-ray systems used with implanted markers as well as implanted radiomarkers, ultrasonic imaging systems and surface alignment imaging systems using lasers and optical patterns. As with patient positional verification, the imaging of the soft tissue with cone beam CT offers an effective means to localize the target. Transit dosimetry (Section 11.12.3) also enables the adaptation of the treatment to movement, providing a cumulative assessment of the dose to the target, the surrounding normal tissue and critical organs through the treatment.

The crucial aspect to positional tracking or the full transit dosimetry tracking is making decisions about changing or modifying the treatment and implementing these. At present changes in bony landmark positioning is often being taken by the machine operators just before irradiation based upon set criteria. Clinical intervention is impractical and intelligent automatic decision systems will be required.

11.13.3 Intra-fractional motion

Adapting treatment for intra-fractional motion remains a challenging prospect for treatment delivery, although it would enable tighter margins to be applied with confidence. Real time tracking of the tumour is required and this is currently being achieved through stereo-radiography of implanted markers or bony landmarks and surface marker tracking. For the latter the need for confident correlation with the internal anatomy is paramount.

The clinical application for this lies in and near the thorax. Primarily lung tumours in the lower lobes, particularly those close to the spinal cord, as well as those affected by diaphragm motion.

For the treatment delivery systems the remaining challenge is to compensate for this motion. Solutions in development are to move the beam limiting device, typically the collimator leafs but also the entire machine in the case of robotic systems. The patient can also be moved using the couch while the beam remains static. Time delays due to mechanical, electrical and computing aspects of the motion compensation system coupled with those of the motion detection system are challenges which need to be overcome in order to deliver accurate four dimensional radiotherapy treatment.

Chapter 12

Brachytherapy

P Bownes, C Richardson and C Lee

12.1 Introduction to principles of clinical use

This chapter should be read in conjunction with Radiotherapy in Practice: Brachytherapy (2nd edition).

Brachytherapy means 'near treatment'. In brachytherapy, the dose distribution is very high close to the source but then falls off rapidly following the inverse square law (see Chapter 10). Brachytherapy is only a practical option if you can access the tumour. It uses radioactive sources placed in a variety of ways:

- Intracavitary—applicators placed in natural body cavities e.g. the uterine canal or vagina.
- Interstitial—needles or catheters directly through the tumour or tumour bed, e.g. prostate or head and neck.
- Intravascular insertion of brachytherapy applicators and sources into an artery to prevent re-stenosis.
- Intralumenal—catheters passed down natural 'tubes', e.g. bronchus or oesophagus.
- Surface moulds/applicators—to treat skin lesions.

Sources can either be put in on a temporary basis until the desired prescription dose is delivered, or permanently, so the dose is delivered over the lifetime of the source. Brachytherapy can either be used as monotherapy or together with external beam radiotherapy.

When brachytherapy started radioactive sources were placed by hand, and staff would receive radiation doses both as they placed and removed the sources but also as patients were cared for during their treatment. Newer technology allowed sources to be loaded after the applicators (specialist catheters to guide the radioactive source) were placed, first manually and then remotely under computer control.

Types of treatment and clinical techniques also vary according to the dose rate:

- Low dose rate (LDR) <2 Gy per hour
- Medium dose rate (MDR) 2–12 Gy per hour
- High dose rate (HDR) >12 Gy per hour

12.2 Source specification and dosimetry

12.2.1 Historical formalisms

Various methods have been used to describe the dose from a source, dating back to when brachytherapy originated. The most commonly used are described in detail in BIR/IPSM recommendations for brachytherapy and also covered extensively in other textbooks. Care should be made when reconstructing patients' treatments that the correct algorithm and data is used. Currently routine brachytherapy planning does not use CT data to take account of tissue inhomogeneities or applicator inhomogenities in the dose algorithm. All doses are computed assuming the medium is water.

12.2.2 Current dose formalism—TG43 Update 1

In 1995 the report of the American Association of Physicists in Medicine (AAPM) Radiation Therapy Committee Task Group 43 (TG43) was published and updated to TG-43 U1 in 2004. It is TG43 U1 formalism which is now used in commercial planning systems.

TG43 U1 is based on measured or measurable quantities (TLD or Monte Carlo generated) produced by a source in a water equivalent medium. The equation is complex but more information is found in Chapter 13 of this book and also in the Appendix of Chapter 3 in Radiotherapy in Practice: Brachytherapy.

12.3 The Paris system (for interstitial treatments)

The Paris system in interstitial radiation therapy was developed to standardize the placement and dosimetry of iridium wire and hairpin implants. It is based on a set of implant distribution rules and then specifies how the dose calculations are to be made. These give low dose rate treatments, typically of 0.5Gy per hour to the prescribed reference dose. The system can also be adapted to afterloader techniques using high activity iridium sources and the placement rules remain valid to achieve a good even distribution of dose.

12.3.1 Source placement

Wires or catheters are implanted through or near the area to be treated and their placement should be according to the following rules:

+ Sources should be straight and parallel.
+ Sources should be of equal length.
+ There should be an equal separation between sources. The separation should be between 5 and 20mm and depends on the volume to be treated.
+ Sources must have equal linear activity.

12.3.2 How do we do a Paris calculation?

+ Dosimetry is calculated on the central plane (Fig. 12.2). The central plane must be perpendicular to the sources midway along the sources. If the sources are not of equal length the central plane should be placed as close to the mean mid length of the wires as possible, where the dose rate between the wires due to their length contribution will be at its maximum. If the wires (or catheters) are not completely parallel the central plane should be perpendicular to the main direction of the source lines, and through the estimated centre of the implant. For hairpin implants the central plane should be half way down the legs of the hairpin, ignoring the crosspiece and orthogonal to the legs.
+ Modern computer systems allow for rotation of the reconstructed implant in three dimensions to easily visualize the implant and choose the calculation plane(s).

Placement of the central plane for a complicated 3D curved implant may be more subjective. The calculation should be discussed with the clinician so that the dose to the treatment volume and organs at risk is calculated appropriately.

◆ Define a set of basal dose points on the central plane—these should be located at the points of minimum dose rate between the wires (Fig. 12.1 & 12.2). The basal points can be defined geometrically. For a single plane the minimum dose will be midway between each pair of wires. For a triangular arrangement of wires the basal dose rates are calculated at the centre of gravity (centroid) of each triangle (the intersection of perpendicular bisectors of the sides of the triangles), and for a square geometrical arrangement the basal dose rates are calculated at the centre of each square.

◆ Calculate the dose rate at each basal dose point.

◆ Calculate the arithmetic mean of the individual basal point dose rates.

◆ Calculate the reference dose rate—this is defined as 85% of the mean basal dose rate. This value was chosen to give an acceptable compromise between a steep dose gradient, whilst giving a reasonable contour coverage of the volume required. Hence the treatment volume is defined as the volume enclosed by the 85% reference isodose.

◆ The reference dose rate can then be used to calculate the treatment time required to deliver the prescribed dose.

When using the true Paris system, only dose points on the central plane should be used. However, modern planning computers can provide dose points and isodoses on multiple planes which can be useful to highlight hotspots or cold areas, in conjunction with isodose distributions. This may be particularly important if carrying out calculations from orthogonal radiograph reconstructions since the soft tissues cannot be indicated in the calculations and so no dose volume histograms will be available.

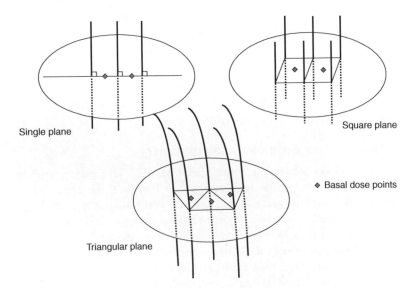

Fig. 12.1 Geometry of Paris implants and basal dose points.
Reproduced from Hoskin and Coyle, *Radiotherapy in Practice: Brachytherapy, Second Edition*, 2011, p. 31 with permission of Oxford University Press.

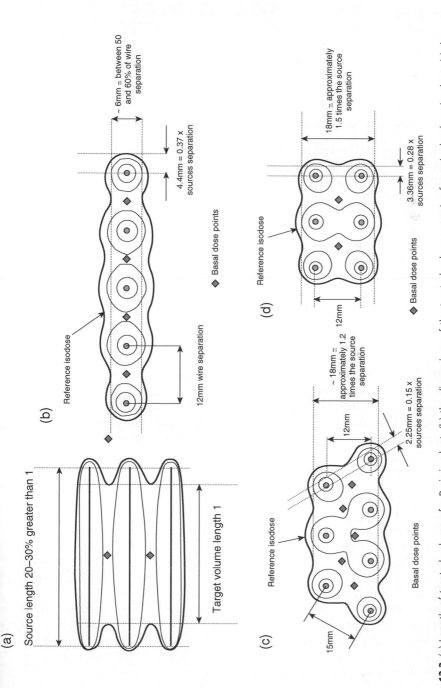

Fig. 12.2 (a) Length of treated volume of a Paris implant, (b) the dimensions of the treated volume margins for a single plane implant, (c) the dimensions of the treated volume margins for a triangular implant, (d) the dimensions of the treated volume margins for a square implant. Reproduced from Hoskin and Coyle, *Radiotherapy in Practice: Brachytherapy, Second Edition*, 2011, p. 33 with permission of Oxford University Press.

Table 12.1 Thickness of the treatment volume

Type of implant	Thickness of treatment volume
Single plane implant.	Between 50% and 60% of wire separation.
Two plane triangular implant.	1.2 x source separation.
Square arrangement.	1.5 x source separation.

12.3.3 Dimensions covered by the treated volume

Since the isodoses will 'pull in' between the wires we need to allow extra coverage to give sufficient margins for the treatment (Fig. 12.2). The length of the treated volume is approximately 0.65 times the length of the sources. This relationship is approximate and depends a little on the wire separation, being relatively smaller for shorter wires so the sources should be about 20–30% longer, at each end, than the target volume. The thickness of the treated volume is indicated in Table 12.1 and Fig. 12.2.

The lateral margins also vary depending on the source separation and are best seen in diagrammatical form. Again they are dependent on the source separation and geometry (Fig. 12.2).

12.3.4 Calculation of decay

When performing dose calculations, the source strength is used, which is normally expressed in terms of a reference air kerma (kinetic energy release per unit mass) on a specified date and time. Modern computer systems will use the activity of the source to calculate the dose delivered but also has to take into account the decay of the iridium source whilst the implant is in place. If not, or when doing a manual calculation using tables it will be necessary to correct for this decay. The Paris literature has a correction table, which adds hours depending on the treatment time. An alternative is to carry out the calculation for the strength of the wire on the middle day of the implant.

12.3.5 Paris calculation examples

12.3.5.1 Example 1

A single plane iridium wire implant consists of four wires, (straight and parallel), each 50mm long, and separation 15mm. Wire strength (at mid implant), AKR, = 450 nGy h^{-1}mm^{-1} at 1 metre.

For a single plane implant the basal dose points are placed midway between the wires and in this case we have three basal dose points between the four wires. Looking at each basal dose point, the distance is measured from that point to each of the four wires. The same is done for each basal dose point in turn. Using a crossline curve or dose rate table for the correct wire length, the dose rate contribution from each wire to each point can be looked up, and thus the dose rate at each point can be calculated. Crossline graphs (or tables) may give the dose rates for wire of AKR 1 μGy h^{-1}mm^{-1}m^2, so the dose rates are then scaled for the actual strength of wire used.

For this example the dose rates at the three points are, 0.395, 0.417 and 0.395 Gyh^{-1}. Therefore the mean basal dose rate is 0.402 Gyh^{-1}. To calculate the reference dose rate the mean basal dose rate is multiplied by 0.85 giving in this case a reference dose rate of 0.342 Gyh^{-1}, so the time required to give a treatment of 65Gy is 190 hours.

12.3.5.2 Example 2

A two plane implant, with five wires arranged in a triangular cross section (as in the triangular plane diagram in Fig. 12.2). Each wire is 70mm long, and the separation is 20mm. We need to calculate the air kerma rate of iridium wire required to give 25Gy in 2.5 days.

Using trigonometry we can calculate the distances from each wire to each point. For example:

From P1 to Wire 1, 3 and 4, 10/x = cos30°, therefore x =10/cos30° = 12.55mm; from P2 to wire 3 and 5, 20/y =cos30° therefore 20/cos30°=23.1mm.

Table 12.2 Dose rates (DR) for 1 μGy h^{-1}mm^{-1} at 1 metre

	Point 1		Point 2		Point 3	
	Dist (mm)	DR	Dist (mm)	DR	Dist (mm)	DR
Wire 1	11.55	0.24	11.55	0.24	23.1	0.095
Wire 2	23.1	0.095	11.55	0.24	11.55	0.24
Wire 3	11.55	0.24	23.1	0.095	30.5	0.062
Wire 4	11.55	0.24	11.55	0.24	11.55	0.24
Wire 5	30.5	0.062	23.1	0.095	11.55	0.24
Totals		0.877		0.877		0.877
Mean basal dose rate			0.888 Gy h^{-1}			
Reference dose rate			= 0.85×0.888 = 0.417 Gy h^{-1}			

A table can then be constructed of the dose rates for a wire strength of 1μGyh^{-1}mm^{-1} at 1 metre.

However, we wish to give 25Gy in 2.5 days (60 hours), so we need to correct for the actual activity of the wire we want to order.

The required reference dose rate is 25/60 or 0.417 Gy h^{-1}. Therefore the required AKR of wire is 1x0.417/0.7548 or 0.552μGyh^{-1}mm^{-1} at 1 metre.

12.4 Gynaecological brachytherapy systems

12.4.1 Manchester system for gynaecological brachytherapy

The Manchester system was developed for gynaecological intracavitary treatments using radium 226 tubes in the 1930s, and the system published in 1938, and later updated, to improve on earlier techniques. The system was designed to standardize treatments, define reference points at which the dose is to be specified, and define source strengths to give a predictable and (almost) constant dose rate to the reference point. Although designed for use with radium tubes, it was adapted for use with caesium 137 tubes in the 1970s and now forms the basis for gynaecological afterloading techniques.

The original Manchester system applicators used a single intrauterine applicator (length 20mm, 40mm, 60mm available) and two vaginal source applicators (ovoids— 20mm, 25mm, 30mm in diameter, all 30mm long). Activity loadings were specified for each size of uterine tube and ovoid. The uterine tube length and ovoid size were chosen to fit the patient and then the whole insertion packed tightly with radio-opaque packing to prevent movement. The standard source loadings gave a dose rate to the defined reference point (Point A) of 54.5 cGy h^{-1} to 55.3 cGy h^{-1}. Vaginal sources should not contribute more than a third of the dose rate to Point A. This dosimetry system was calculated assuming that the applicators were perfectly inserted and totally symmetrical.

Point A was originally defined as being 2cm lateral to the centre of the uterine canal and 2cm superior to the mucous membrane of the lateral fornix along the line of the uterine canal. The dose at Point A was chosen to be representative of the minimum dose to most of the malignant tissue when treating cancer of the cervix. An additional calculation Point B, was defined to estimate the dose to the pelvic wall. Point B was placed 5cm laterally to the patient's midline and at the same level as Point A. When the system was first implemented, the calculations were all carried out for ideal insertions and no individual patient calculations were carried out.

To accommodate changes in technology many users reinterpreted the definitions of the Manchester system. The definition of Point A was adapted so that the vaginal mucosa was assumed to be at the level of the cervix—practically this meant the two Points A are 2cm superior to the bottom of the uterine tube sources, along the line of the tubes, and 2cm orthogonal to the tube, to the patient's right and left (Fig 12.3).

To compare patients' treatments with those of another centre, detailed information is required as to the definition of the reference points, and the doses received by them, the target volume and the surrounding tissue. These details are fully reported in the ICRU Report 38 recommendations and rectal and bladder points are shown in Fig. 12.3.

12.4.2 Use of Manchester system for afterloading systems

A variety of gynaecological applicators are available for use with afterloaders The advantage of using such a set of applicators is that the applicators can be fixed in a rigid geometry, so that standard treatments can be given according to the Manchester system. New applicator systems such as uterine tube and vaginal ring systems, where the ring mimics the ovoids, have also been introduced. Both ring systems and ovoid systems have been developed with additional holes to allow interstitial needles to be

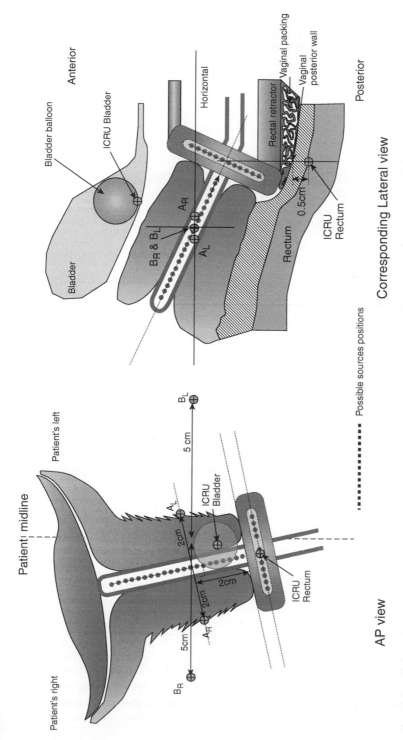

Fig. 12.3 Manchester system applied to Ir-192 afterloader ring applicator. Reproduced from Hoskin and Coyle, *Radiotherapy in Practice: Brachytherapy, Second Edition*, 2011, p. 27 with permission of Oxford University Press.

placed for better coverage of large or asymmetric tumours. Thus gynaecological treatments can now be both intracavitary and interstitial.

The loading of the afterloader source trains, or planning of a stepping source can be arranged to mimic the Manchester source loadings. Thus the traditional clinical treatment can be given whilst utilizing the radiation protection advantages of afterloaders. The uterine tubes can still be used to define Points A and B, since the location of the tubes relative to the cervix is known. The only adjustments may be when using a ring system, since there is no flange to delineate the bottom of the cervix. In this case the location of Point A(s) is 2cm superior to surface of the ring and 2cm lateral to the uterine tube.

12.4.3 Three dimensional image guided brachytherapy for cervix cancer

Over the last ten years there have been considerable advances in computer software which allows utilization of optimization, 3D evaluation of isodoses, including, dose volume histograms (DVH) and image fusion (see Chapter 10 for further details). Many applicators are now CT and MR compatible and so it is appropriate to reconsider how gynaecological treatments are planned and evaluated.

Concepts for 3D imaged based treatment planning of the cervix brachytherapy have been developed and are detailed in the GEC-ESTRO recommendations. The RCR have also published guidelines for implementation in the UK. They suggest the following:

- ◆ CT/MRI compatible applicators.
- ◆ CT/MRI imaging and fusion.
- ◆ Define the following target volumes using MRI.
  - • GTV_B—macroscopic tumour at the time of brachytherapy.
  - • High risk clinical target volume (HRCTV)—includes GTV_B + whole cervix + presumed extra cervical tumour extension.
  - • Intermediate clinical target volume (IRCTV)—is based on the macrosopic tumour extension at diagnosis.
- ◆ Organs at risk (OAR) to be marked on: bladder, rectum, sigmoid, small bowel.
- ◆ Dosimetry reported for volume based image guided brachytherapy.
  - • Concept of the equivalent dose in 2Gy fractions (EQD2) is recommended, along with physical dose.
  - • The minimum dose to 90% and 100% of the HRCTV, $Gy_{\alpha/\beta10}$ (D90 and D100 respectively). The volume receiving 100% of the prescription dose is also a useful parameter.
  - • The minimum dose to the most exposed 2cm^3 and 0.1cm^3 of the OAR, $Gy_{\alpha/\beta3}$ (D_{2cc} and $D_{0.1cc}$ respectively).
- ◆ Total reference air kerma (TRAK), Point A doses, ICRU reference points for rectum and bladder and description of the time dose pattern should be reported.

In practice, the Point A prescriptions may often still be appropriate, but the possibility to optimize treatments and conform the isodoses allows both better coverage of the tumour volume and reduced doses to organs at risk. It also allows the opportunity for dose escalation with MRI based brachytherapy with the potential to improve local control without increasing toxicity.

12.4.4 Endometrial and vaginal treatments

In addition to the cervix applicator sets there are also a range of applicators available to treat both the endometrium and vaginal vault, and combinations of these. Centres may have their own systems to plan these but the principles of evaluation remain the same as for cervical treatments. For vaginal vault treatments the applicator used is a single line source encompassed within a cylinder (varying diameter to fit patient) to give surface dose sparing. The dose is traditionally prescribed to 5mm from the applicator surface. If the prescription depth is increased it is important to evaluate the surface dose as well. Alternative applicators exist for more complex treatments, e.g. partially blocked vaginal applicators, or applicators with needles embedded in them, such as the Nucletron Miami applicator. These applicators, when used with optimized planning, allow reduction of the doses to organs at risk and allow higher dose to certain areas.

The clinical application of brachytherapy is covered in detail in the other text, Radiotherapy in Practice: Brachytherapy. A brief summary of the main areas of use follows.

12.5 Prostate brachytherapy

Brachytherapy is ideal for the treatment of prostate cancer where the target volume lies very close to critical normal tissue, in particular the anterior rectal wall and bladder. Both permanent seed brachytherapy (Iodine-125 or Palladium-103) and temporary high dose rate (iridium-192) implants are now mainstream prostate cancer treatments.

The efficacy and success of both techniques have been improved by appropriate patient selection (refer to relevant GEC-ESTRO recommendations), image guidance and advances in treatment planning. The predominant image guidance technique is transrectal ultrasound (TRUS) imaging coupled with a template that allows real time image guidance and accurate source placement via a transperineal route.

Treatment planning for seed brachytherapy is based on TRUS imaging directly linked to the planning system, whilst the patient is in the lithotomy position. For HDR prostate brachytherapy planning can be based on TRUS imaging or CT/MR imaging. The following volumes should be defined:

- Gross tumour volume (GTV) is contoured whenever possible.
- Clinical target volume (CTV) is the prostate gland, (for HDR may include any appropriate seminal vesicle or extra capsular involvement), plus a 3mm margin in each. This can be constrained to the rectum posteriorly and bladder neck cranially.
- It is recommended that the planning target volume (PTV) = CTV as there are no significant set up errors and real time image guidance is used.

- Organs at risk are rectum, prostatic urethra and bladder base.

The core planning aims which effect the seed placement or the HDR dwell position/dwell time are:

- Conformity of the prescription isodose to the CTV.
- Maintain acceptable dose homogeneity.
- Minimize the dose to the organs at risk.
  - For the rectum the minimum dose in the most irradiated 2cc volume of the rectum (D_{2cc}).
  - For the prostatic urethra the minimum dose in the most irradiated 10% of the prostatic urethra (D_{10}).

The recommended monotherapy prescription dose for permanent seed brachytherapy is the intended dose to the 100% isodose and is 145Gy for Iodine-125 seeds.

Most common HDR boost fractionation regime currently in the UK is, 15Gy in a single fraction to CTV followed by external beam radiotherapy of 37.5Gy in 15 fractions.

Source or applicator implantation is performed under real time TRUS image guidance and the utilization of stepping unit. The applicator position or seed position can be dynamically fedback to the treatment planning system and the plan adapted if required.

For permanent seed brachytherapy it is essential to perform post implant dosimetry to all permanent seed brachytherapy procedures and this allows you to:

- Determine the actual dose delivered.
- Investigate relationships between planned and achieved dosimetry.
- Correlate actual dosimetric parameters to clinical outcome.
- Improve technique and clinical efficacy.

Due to the remit of this book the authors were unable to include details on important guidance regarding patient selection, equipment used, volume definition, treatment planning, uncertainties in treatment delivery and relevant QA issues. For this information please refer to the relevant GEC-ESTRO recommendations and the Radiotherapy in Practice: Brachytherapy book.

12.5.1 **Breast brachytherapy**

Brachytherapy to the breast is most commonly delivered as a boost following a course of external beam radiotherapy. Brachytherapy allows the delivery of a high dose to the tumour situated within the breast while maintaining sparing of the skin and underlying structures.

There is an increasing body of evidence that suggests that accelerated partial breast irradiation, where the treatment volume is the primary surgical site alone plus a suitable margin, may be appropriate for early stage breast cancer. Brachytherapy can be used as the sole modality for this treatment. The dose is usually delivered with a high dose rate stepping source using a multicatheter intersitial implant or a balloon catheter such as the Mammosite®.

12.5.2 **High dose rate rectal brachytherapy**

Rectal cancers are usually removed by surgery. However, to make surgical removal easier and also to preserve the anus, HDR brachytherapy can be carried out before surgery. For patients who are unsuitable for surgery, rectal brachytherapy can also be given in addition to other forms of non-invasive treatment.

12.5.3 **Head and neck brachytherapy**

Traditionally head and neck brachytherapy boosts have been given using iridium wire which is sealed in a 4F diameter inner tubing and then manually afterloaded into 6F outer plastic catheter inserted through the required area either at the time of a neck dissection or as required. The catheters are placed according to the Paris system and the typical dose rate is 0.5Gyh. These cases can be treated using a micro HDR afterloader. Commercially available systems allow inner tubing to be inserted into needles or outer tubing in the same way. The treatment is then fractionated to give the appropriate dose—for example 42Gy in 14 fractions over 10 days.

Treatments using Ir-192 'hairpins' were often used for small tumours of the oral cavity. Hairpins are no longer commercially available, so these treatments can be given with manually constructed hairpins, or with an afterloader system. Another applicator, the Nucletron Rotterdam applicator, used to treat recurrences in the nasopharynx, is a specialized form of surface mould. This consists of two approximately parallel applicators and a Paris system of basal dose points can be set up to give an initial dose calculation. CT images can be used to delineate the tumour and the dose can be optimized to give the best coverage to the tumour whilst sparing the OAR.

12.5.4 **Surface moulds**

A surface mould is a custom made device which attaches to the patient and supports applicators or radioactive sources at a fixed distance from the skin surface, typically between 5 and 20mm. Traditionally the mould would contain a range of sources but now can be designed to incorporate catheters attached to an afterloading system.

Dosimetry systems such as the Paterson-Parker rules were used to specify the arrangement for sources for the desired treatment area. The choice of distance from the sources to skin surface is dependent on the depth of treatment and the allowed skin dose, since the fall off of dose is dominated by the inverse square law; the closer the source is to the skin the greater the skin dose. The dose to the treated area varies by +10% due to the non-uniform isodoses produced by the spaced, discrete sources.

12.6 **Optimization techniques in brachytherapy**

12.6.1 **Background**

Optimization techniques allow the user to choose the location of the radioactive source, and the length of time it is in that location to alter the dose to the target and to the organs at risk. This can be done manually, or by using one of several computing methods. Often the best technique is to allow the computer to do its best, and then manually 'tweak' the result.

12.6.1.1 Simple case

If all the dwell times for a stepping source are equal for a single line treatment the overall longitudinal shape of the reference isodose will be that of a cigar—broader in the centre of the treatment and narrower at the ends. If a cylindrical treatment is required, by reducing the dwell position times in the centre with respect to the ends the isodoses can be made to conform closer to the cylindrical ideal. For a single line treatment using regularly spaced dwell positions, the dwell times in the centre of an optimized treatment will be between a third and a half of the treatment times at the end positions.

12.6.1.2 Disadvantages of optimization

Care must be taken in altering treatments given in practice. Changing to an optimized technique alters the dosimetry and no longer satisfies the classical dosimetry systems previously in use. The active length of the catheters can be shorter, since the dwell times at the periphery of the volume treated can be increased to improve the coverage, by flattening the reference isodose surface at the outer ends of the catheters.

12.6.2 Optimize on points

Computer packages will allow a variety of points of interest to be added.

- ◆ Basal dose points—can be added by the computer but care should be taken that they conform to the Paris system.
- ◆ Patient points—are manually entered points at which the dose can be calculated, e.g. ICRU bladder and rectal points.
- ◆ Applicator points—can be put on points of interest or at a specified distance from an applicator.

Calculations and prescriptions can be based on any of these but it is important to be clear which method is chosen.

12.6.3 Points on a target

It may be appropriate to outline a target and then put points on its surface. The dose can then be prescribed to the surface—a straight calculation will provide a mean and the optimization package can then attempt to cover the target better, by creating a set of simultaneous equations to solve—be careful to assess organs at risk for hotspots.

12.6.4 Points within a volume

These can be Paris calculation points, patient points of interest or the computer generating a series of dose calculation points within a volume.

12.6.5 DVH analysis

When the dose calculation is complete it can be assessed against other plans and agreed dose limits using. Dose volume histograms (DVHs). These are discussed further in Chapter 10.

12.7 Quality assurance in brachytherapy

12.7.1 Introduction

A quality assurance (QA) programme is required to ensure all treatments are delivered correctly and that the IR(ME)R practitioner's clinical intent is executed safely with regard to the patient and others. Tolerances are set to ensure a high level of safety regarding the patient and staff and acceptable accuracy of the dose delivery in terms of source strength, positional accuracy and temporal accuracy.

A QA programme should address every step of the treatment process:

♦ Diagnosis and treatment decisions,

♦ Implant design and applicator/source insertion process,

♦ Verification of source strength and distribution of activity within source,

♦ Definition of target volumes and normal tissue structures,

♦ Reconstruction of applicator positions,

♦ Treatment planning process,

♦ Treatment delivery process, including positional and temporal accuracy.

IPEM Report 81 provides the UK with recommendations for the quality control required in brachytherapy with additional recommendations given in the BIR/IPSM recommendations for brachytherapy, and ESTRO Booklet 8. Quality assurance issues for afterloading units are also covered extensively in other texts. QA for low dose rate (LDR) and medium dose rate (MDR) remote afterloading units will not be covered as the majority of units in the UK are becoming obsolete.

12.7.2 High dose rate (HDR) remote afterloading

A typical QA schedule for a HDR remote afterloading unit is shown in Table 12.3 and similar tests can be applied to pulse dose rate (PDR) remote afterloading units. The core elements include:

♦ Verification of source air kerma rate,

♦ Positional verification of the source,

♦ Temporal verification,

♦ All machine interlocks and safety features function correctly.

For HDR units sources are replaced every three months. The end user must independently verify the reference air kerma rate of the new source to within ±5% of the calibration certificate of the manufacturer. Measurements must be performed using equipment traceable to a national standard. In the UK the NPL have developed a direct traceability route back to the complex spectrum of iridium-192 using a HDR remote afterloading source.

12.7.3 Manual techniques (manual application and manual afterloading techniques)

For temporary and permanent manual source insertions (I-125, Pd-103, Cs-137, Ir-192 wire etc.) the core QA tests required are shown in Table 12.4:

Table 12.3 Typical quality assurance schedule for a HDR remote afterloading unit

Frequency	Test	Method	Tolerance
Pre-treatment	Machine function tests	Check correct function of all interlocks, indicators, emergency equipment, area radiation monitor, audio and visual systems.	
	Source data checks	Verify date, time and source strength in planning computer and treatment unit.	
	Positional accuracy	Verify source position of the stepping source either using a source stepping viewer and CCTV or with film.	± 1mm
	Temporal accuracy	Check temporal accuracy against a stopwatch. Consistency and linearity should be verified.	± 1%
	Applicator integrity	Check visually for damage and treatment simulation.	
Quarterly (post service and source change)	Source calibration	Two independent checks on the air kerma rate of the new source. Data entry into the treatment planning system and treatment unit should be verified.	± 5%
	Source position set up and verification	Verify the location of the source in its safe position. Verify source position using both a source stepping viewer with CCTV and film.	± 1mm (aim for ± 0.5mm at set up)
	Transit time	Ensure the transit time is consistent and transit dose is negligible.	
	Pre-treatment tests	As above	
Post maintenance		Relevant QA performed depends on what maintenance has been performed.	
As required	Applicator checks	New applicators should be checked using radiographic and autoradiographic techniques to verify applicator integrity and source position within it. Check the marker wires used in treatment planning imaging correlate to the actual source position at treatment.	

12.7.4 Treatment planning systems

Brachytherapy treatment planning systems have increased in complexity so planning system quality control should be designed to ensure the quality of treatment plans, minimize the possibility of systematic errors and be designed to complement checks on individual treatment plans.

The frequency of testing depends to some extent on the brachytherapy workload of the department, but if several calculations a week are made then a typical schedule is

Table 12.4 Typical quality assurance schedule tests for manual brachytherapy techniques

Test	Frequency	Description of test procedure	Tolerance
Verification of source documentation.	Each Source	Ensure correct source is delivered.	
Measurement of source strength.	Pre-use	Use a measurement system traceable to a national standard for that particular source type (energy, source geometry) and source holder (position of source and orientation). A re-entrant well chamber is commonly used.	±5%
Verification of activity distribution within the source.	Pre-use	Autoradiographs should indicate uniformly distributed activity within the source and can also be used to verify the configuration of pre-loaded source chains.	
Leakage testing.	Before first use then annually	Leakage tests consist of wiping the radioactive source using forceps with a swap moistened with water or methanol. The swab is then measured using a sodium iodide scintillation counter.	<200Bq
Applicator integrity.	Pre-use	Visual inspection for damage.	

shown in Table 12.5. For HDR remote afterloading systems the data for each new source must be entered into the planning system and be independently checked by a second physicist.

12.7.5 **Auxiliary equipment**

Routine QA programmes should also cover quality assurance on the measurement equipment used to perform source calibration. Commonly long lived sources are used

Table 12.5 Typical quality assurance schedule tests for brachytherapy treatment planning systems

Frequency	Test	Tolerance
Pre-use	Calculation of a simple standard plan.	No change for same source strength.
	Data file integrity, using check sum methodology.	
Every clinical Use	Consistency of printed plan documentation and data transfer to treatment unit.	
	Independent plan check, including an independent dose calculation.	±5%

(continued)

Table 12.5 (continued)

Frequency	Test	Tolerance
Three monthly checks	More complex dosimetric tests, e.g. multiple source, multiple channels, optimization techniques.	±3%
	Geometric reconstruction of applicators from all imaging modalities used.	mean I deviation I ±1mm
	Image registration techniques.	
	Plan evaluation tools—use of dose volume histograms over a range of scenarios.	DVH Parameters <+/–3%
	Image acquisition. CT and MRI will have QA programmes for the scanners themselves. May need brachytherapy specific tests. The relevant code of practice and equipment to perform these measurements is discussed in detail in Chapter 13 of the Radiotherapy in Practice: Brachytherapy book. Ultrasound in conjunction with a stepping unit device is used for prostate brachytherapy. Specific tests on the volume acquisition, template alignment, spatial accuracy and digital connectivity to the planning system should be checked.	
	Overall system tests should be performed, which test the complete planning process from start to finish to check on the systematic behaviour of the treatment planning process.	±5%

to check stability of response, e.g. strontium-90 checks are used for thimble ionization chambers and caesium-137 sources are used for reentrant well chambers. This is performed at least annually and would expect readings to be within ±1% of baseline. Cross calibration of equipment back to the National Standard (NPL) is currently recommended every 3 years.

Radiation protection monitors should also undergo annual calibration as part of a QA programme or when deemed necessary. Checks on source preparation tools and shielding equipment should be performed pre-use. Tools used to prepare iridium wire should have routine wipe tests performed (<200Bq).

Chapter 13

Radioactive sources

C Richardson, G Workman and P Bownes

13.1 **Introduction**

Radioactive sources can be either naturally occurring, or man-made. The nuclei of these sources are unstable if the number of protons and neutrons in the nuclei is not balanced and the nuclear forces cannot hold the nuclei together. An unstable nucleus will naturally and spontaneously change to achieve a more stable combination. Radioactive decay processes are discussed in Chapter 2.

13. 2 **Requirements for clinical sealed sources**

There are many radioactive sources which have been used for brachytherapy since the use of radium in the early 1900s. By the 1920s designs were available which encapsulated the radium, containing it and filtering out the beta radiation. It was then used extensively throughout the world, but has several disadvantages, so has now been replaced by other man-made isotopes which are shorter lived and have fewer hazards associated with them. Most departments offering brachytherapy will now have an iridium source in an afterloading unit, and use either iodine-125 or palladium-103 for prostate implants. Other types of sources are used for specific purposes, such as eye plaques, which may use beta emitters (e.g. Sr-90). It is still important to be aware of the earlier treatments and sources since the treatments were clinically effective and patients may still be around!

See Table 13.1 for a list of commonly used radioactive sources, including those used historically. Choosing which one to use will depend on a variety of things.

- ◆ Energy of gamma ray emission.
  - High enough to avoid increased energy deposition in bone by the photoelectric effect.
  - High enough to minimize scatter.
  - Low enough to minimize radiation protection requirements.
  - Optimum energy up to 0.4MeV.
- ◆ Half-life
  - Very short half-lives are unsuitable.
  - Long half-life sources are expensive to dispose of.
  - The half-life should be such that correction for decay during treatment is minimal.

Table 13.1 Radioactive sources for clinical use

Source	Form	Production	Half-life	Emissions
Ra^{226}	Cylindrical	Naturally occurring	1620 yr	2.45 MeV (max)
Cs^{137}	Cylindrical tubes spherical pellets needles	Fission product	30 yr	0.662 MeV gamma
Cs^{131}	Seeds	Neutron activation	9.7 days	30.4 keV (mean)
Co^{60}	Cylindrical	Neutron activation	5.26 yr	1.17, 1.34 MeV gammas
Au^{198}	2.5mm grains 14 in magazine	Neutron activation	2.7 days	0.412 MeV gamma
Ir^{192}	Wire, pins hairpins	Neutron activation	74 days	0.38 MeV (mean) gamma
Pd^{103}	4. 5mm seeds with radiographic markers	Neutron activation	17 days	21 keV (mean) X-rays
I^{125}	4.5mm seeds variety of types available	decay product Xe^{125}	59.6days	27.4, 31.3 and 35.5 keV X-rays
Sr^{90}	Eye plaque wire (infused on aluminium)	Fission product	Effective half-life 28.7 years	2.27MeV beta particles
Ru^{106}	Eye plaque	Fission product	Effective half-life 1.02 yr	3.54MeV beta particles

- A long half-life means that permanent stock will have little radioactive decay during the lifetime of the source—this is now only appropriate for afterloaders.
- Cost should be reasonable, both to buy and also to dispose of.
- Specific activity (yield).
  - Should be high so there are many decay events for a small volume of source— easier to fit in an afterloader, and will go down a smaller diameter applicator increasing the number of clinical sites which can be treated.
- Type of decay by-products.
  - Charged particle emission should be absent or easily screened (except for beta emitters if required).
  - There should be no gaseous disintegration product.
- Physical form
  - Available in insoluble and non-toxic form.
  - The material should not powder or be dispersed if the source is damaged or incinerated.
  - Damage during sterilization should not be possible.

- The isotope should be able to be made in different shapes and sizes, e.g. rigid tubes, needles, small spheres and flexible wires.

Quite a list! But in practice for most brachytherapy the choice of source goes hand in hand with the choice of technique.

13.3 Production of clinical sources

Radionuclides for brachytherapy can be produced by either fission or neutron activation (bombardment) in a nuclear reactor.

Fission products occur when large nuclide divide and produce new elements which are radioactive and will have to be separated out from any other fission products. The main example of this is Caesium-137.

In **neutron activation**, (the 'n-γ' reaction) a sample of a stable isotope is placed in a neutron field in a reactor. Some of the nuclei of the element capture a neutron and become a radioactive isotope of that element, with the emission of gamma radiation. The result will be a combination of the stable and the radioactive isotope. Many modern brachytherapy sources are produced like this.

Sealed sources, used in brachytherapy, are solid, discrete and encapsulated. Hazards are linked to security and external radiation dose rather than contamination. See Chapter 15 on Radiation Protection. See Table 13.1 for the physics details for the sources.

It should be noted that radium-226 and gold-198 are no longer available. Details on their decay schemes can be found in Radiotherapy in Practice: Brachytherapy, Chapter 2. Caesium-137 is currently being phased out of use.

13.3.1 Caesium-137

Caesium-137 became available in the late 1960s and rapidly replaced radium-226. It is a product of uranium fission and decays by beta minus emission. It has a half-life of 30.17 years and emits a photon energy of 0.662MeV.

$$^{137}_{55}Cs \longrightarrow\ ^{137}_{56}Ba + \ ^{0}_{-1}e + \gamma$$

Since this is a lower photon energy than that of radium-226 radiation protection was made easier, and since its daughter products are solid it was less hazardous if damaged.

Caesium tubes were designed to replace radium tubes for gynaecological treatments and so the Manchester system was directly transferable from one radioisotope to the next. Caesium spherical sources were also used in afterloaders. In both tubes and pellets the caesium is incorporated onto beads and then encapsulated in stainless steel.

13.3.2 Cobalt-60

Cobalt-60 is produced by neutron activation of cobalt-59 and has a half-life of 5.26 years and decays by beta minus emission, producing gamma energies of 1.17 and 1.33MeV.

$$^{60}_{27}Co \longrightarrow\ ^{60}_{28}Ni + \ ^{0}_{-1}e + 2\gamma$$

Cobalt-60 was used for brachytherapy in a variety of forms such as tubes, needles and also in the form of pellets for high dose rate afterloading machines. The pellets are similar in design to the caesium pellets used in LDR/MDR afterloading systems. Their shorter half-life required a more frequent change in sources, and as with the caesium systems the relatively large diameter of the pellets restricted the use of the afterloaders. In addition the relatively high energy of the gamma radiation required more shielding than that required by caesium-137.

13.3.3 **Iridium-192**

Iridium-192 is produced by neutron activation of iridium-191. It has a half-life of 73.83 days and decays by beta minus emission.

$$^{192}_{77}Ir \longrightarrow {}^{192}_{78}Pt + {}^{0}_{-1}e + \gamma$$

The photon spectrum is complex with a weighted mean of about 0.38MeV. Iridium is commonly used in two forms, wire which can be cut to length by the user, and as a high activity source in an afterloader.

Iridium-192 wire is constructed with an inner core of a radioactive iridium/platinum alloy. This is surrounded and contained by a 0.1mm thick platinum sheath which also shields the beta minus particles which are not required.

The source is delivered as set lengths of wire but then needs to be cut to the required length. The platinum sheath will continue to seal the source around but not at the ends so it is possible for the core of radioactive material to slide out. The cut iridium wire needs to be encapsulated before clinical use—usually in purpose designed plastic tubing which is heat sealed to contain the source.

13.3.4 **Iodine-125**

Iodine-125 is used principally for prostate brachytherapy. It is a daughter product of xenon-124. Iodine-125 decays by electron capture, emitting a gamma ray of 35.5keV and characteristic radiation of 27.4 and 31.4keV. Iodine-125 has a half-life of 59.4 days.

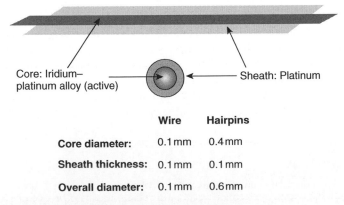

Core: Iridium–platinum alloy (active)   Sheath: Platinum

	Wire	Hairpins
Core diameter:	0.1mm	0.4mm
Sheath thickness:	0.1mm	0.1mm
Overall diameter:	0.1mm	0.6mm

Fig. 13.1 Diagrammatic representation of Iridium-192 wire.
Reproduced from Hoskin and Coyle, *Radiotherapy in Practice: Brachytherapy, Second Edition*, 2011, p. 9 with permission of Oxford University Press.

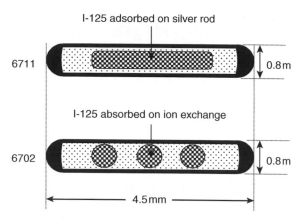

Fig. 13.2a Two I-125 seed designs manufactured by Oncura.

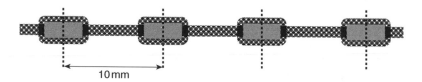

Fig. 13.2b Diagrammatical representation of I-125 seed sutured strands.

$$^{124}_{54}Xe(n,\gamma)^{125}_{54}Xe \xrightarrow{\beta^-} \,^{125}_{53}I + \,^{0}_{-1}e \xrightarrow{ec+x-rays} \,^{125}_{52}Te + \gamma$$

Such a low energy radiation means that shielding is easy, and radiation concerns post implant are confined to those regarding losing sources, both during surgery or post-mortem. For brachytherapy the I-125 is usually incorporated into implantable seeds, which come in a variety of constructions. Some seeds contain radiographic markers but others do not. As implantation techniques alter, and ultrasound is now usually used to guide implantation, it is important that the seeds can be visualized using the modality required. A commonly used type is the Oncura type 6711, being a typical size of 4.5mm long and 0.8mm diameter. Seeds are often stranded together using dissolvable suture for ease of implantation.

The dosimetry of each type of seed will be individual and care should be taken to use the correct data when planning implants. See TG43 U1 for the most comprehensive dosimetry data.

In addition I-125 is also sometimes used in eye plaques.

13.3.5 **Palladium-103**

Palladium-103 decays by electron capture with a half-life of 17 days, and like iodine-125, it emits a mixture of gamma rays and characteristic radiation, producing a slightly lower mean energy of about 21keV. It is encapsulated into seeds of the same dimension as iodine-125 seeds and used in a similar manner.

$$^{103}_{46}Pd + ^{0}_{-1}e \longrightarrow ^{103}_{45}Rh + \gamma$$

Palladium-103 can also be used in eye plaques.

13.3.6 **Strontium-90**

Strontium-90 is a fission product and is used in brachytherapy as a beta emitter for superficial treatments. It decays by beta minus emission with a half-life of 28.7 years to yttrium-90.

$$^{90}_{38}Sr \rightarrow ^{90}_{39}Y + ^{0}_{-1}e$$

The beta energy from strontium-90 is only 546keV and is therefore not useful for brachytherapy but the daughter product of strontium-90 is yttrium-90 which only has a half-life of 64 hours but decays by beta minus emission of a maximum energy of 2.27MeV.

$$^{90}_{39}Y \longrightarrow ^{90}_{38}Zr + ^{0}_{-1}e$$

The combination of strontium-90 and yttrium-90 therefore provides a source with a practical energy of 2.27MeV(E_{max}) and a half-life of 28.7 years.

Strontium-90 was incorporated in a variety of designs of surface applicators, particularly ophthalmic applicators and where they could give a high surface dose with a rapid fall off beneath. Treatments required holding the applicator against the area required for a few minutes. The use of strontium-90 eye plaques may be waning but new type of handheld afterloader has been developed for a clinical trial treating wet macular degeneration of the retina. This contains an aluminium insert infused with strontium-90. The source is welded onto a stainless steel wire and contained within a tungsten alloy for shielding. Delivery of the source is via a cannula.

13.3.7 **Ruthenium-106**

Ruthenium-106 has largely replaced strontium-90 for eye plaques. It also emits beta minus radiation with higher maximum energy of 3.54 MeV and has a half-life of 374 days.

$$^{106}_{44}Ru \longrightarrow ^{106}_{45}Rh + ^{0}_{-1}e$$

The plaques consist of a thin film of Ru-106, encapsulated in pure silver, with a thin 'window' on the concave side to allow treatment, and a thicker backing shield for radiation protection on the convex side. The plaques are shaped to fit the eye, with a variety of radii, and with suture eyelets to allow fixation. Used in ophthalmic plaques to treat uveal melanoma, retinoblastoma, melanoma of the iris and other tumours.

13.3.8 **Other radionuclides**

Ytterbium-169 has a mean photon energy of 93 keV, a half-life 32 days and is being used for breast and prostate implants.

Caesium-131 has a mean photon energy of 30.4 keV, a half life 9.7 days and is being investigated for several implant sites—the shorter half-life may be a big advantage, giving a higher biologically effective dose, and the slightly higher energy may reduce inhomogeneity but long term toxicity data is not yet available.

13.4 Afterloaders

To reduce dose to staff and allow time for more accurate positioning and improved geometry, afterloading systems were developed to allow the applicators to be positioned in the patient and then the radioactive sources introduced later.

Manual afterloading systems were developed for gynaecological applicators and Iridium wire systems, which reduced doses but did not remove the hazard. To eliminate doses to staff completely computer controlled systems are required. There have been a wide variety of designs used but essentially they fall into two categories. They are either a pellet system (or source train) or a stepping source fixed on a drive cable.

13.4.1 Pellet afterloader systems

In a **pellet system** source pellets and spacers are programmed and assembled to make a source train which can be composed differently for each treatment. The applicators and the afterloading machine form a closed system when they are attached. A positive air pressure system forces the pellets from the safe into the applicators. Pellet systems could be either low, medium or high dose rate. A common example of this was the nucletron Cs-137 MDR/LDR system, which used 2.5mm diameter pellets.

The source positions and catheter times were programmable and for cervix treatments the sources would stay in the patient for between 10 and 20 hours. Each time the patient needed care the treatment could be interrupted, but of course this lengthened the overall treatment time.

Another disadvantage of this system is that all the pellets within an individual catheter have to remain in that catheter for the same time, so the system is not particularly flexible. The relatively large diameter of the pellets limited the variety of tumours which could be treated and this type of afterloader is not suitable for interstitial work.

13.4.2 Stepping source systems

Originally the **stepping source** system catheters were also relatively large in diameter, but the most recent advances allow the use of micro afterloader, where the active source is an iridium-192 source less than 1 mm in diameter, which can fit in much smaller catheter delivery systems. Nucletron's HDR MicroSelectron or Varian's Gammamed are examples of such systems. A single stepping source system moves the source to pre-determined positions using a stepper motor/cable system. Although there is only one source, the position and dwell time of the source are programmable so the optimal location and time can be chosen. The only disadvantage is that since there is only one source the treatment times can be significant as the source reaches the end of its clinical life. An iridium-192 source has a half-life of 73.83 days and so is usually changed approximately every three months. This type of afterloader can be a high dose rate system, but can also be adapted to use a lower activity source to deliver pulsed treatments to mimic low dose rate iridium wire or caesium treatments.

Essential features of an afterloader:

◆ Multiple channels can be connected to applicators,

◆ A variety of applicators available,

◆ Adapters available to allow easy source activity checks & QC checks,

◆ Thin flexible source to go around tight curves in applicators,

◆ Easily programmable treatment console,

◆ Direct plan transfer from planning system to treatment machine,

◆ Easy to use!

Treatment machine—safety features:

◆ Back up secondary timer system,

◆ An automatic check of the transfer tube/catheter system before the source is exposed,

◆ Built in source position(s) checks,

◆ Operating system to check that the sources have returned properly,

◆ Back up power supply,

◆ Source held in a safe so low dose around machine when not in use,

◆ Manual source return in the event of complete power failure,

◆ Automatic retention of treatment data and history in the event of power failure,

◆ Alarm and status code system to alert user to faults. etc.

13.4.3 Afterloading machine QA

An afterloader will be used repeatedly so it is essential that it works correctly. Some form of QC check needs to be carried out every treatment day to ensure the software and mechanics are still in working order, but other checks may only be needed after the radioactive source has been exchanged. The regularity of checks may depend in part on the number of patients treated per week. If only one treatment session or patient is scheduled then all the checks should be carried out as pre-treatment checks.

A schedule of tests are required and thought should be given to the following :

◆ Frequency—when, and how often, the test should be performed.

◆ Description—a description of the procedure to be followed.

◆ Results—the expected values or results of the test, including acceptable ranges when appropriate.

◆ Documentation—a form to be completed, dated and signed when the test has been carried out.

13.5 Hazards with sealed sources

13.5.1 Control and testing of sealed sources

For more details on handling and storage of radioactive sources, and on brachytherapy treatment room design, see Chapter 15 (15.7.6.9) on radiation protection. It is very important that all sources are kept secure, sufficiently shielded and their whereabouts known at all times.

Any department handling manually loaded sources, including Ir-192 wires, Cs-137 tubes and I-125 seeds will need systems for QC and auditing which include the following as required by the legislation.

1. Calibration—on receipt.

2. Safe audits (stock checks)—at monthly intervals.

3. Wipe tests—annually.

4. Tally board—current status/location of all sources.

5. Documentation—lots!!

6. Disposal—of sources no longer required.

13.5.2 Storage and movement control

A range of paperwork is required to ensure that all radioactive materials are accounted for. Details of the sources need to go with the sources at all times, so when sources are implanted in a patient the paperwork should follow the patient, giving details of the type and number of sources, the total activity and reference date, and who is 'in charge' of the sources—whether it is the source office technician, whilst in theatre, or the nurse in charge of the patient on the ward.

When the patient is on the ward equipment needs to be available to retrieve any active sources which may become dislodged. The ward should have their own hand held monitor to identify lost sources, and storage pots and handling forceps in case they have to retrieve them.

13.6 Specification of source strength

13.6.1 Activity content

The concept of activity content is used if the source output can also be quantified by the amount of radionuclide within a source. The activity of an amount of radioactive nuclide is defined by

$$A = dN/dt \tag{1}$$

where dN is the expected value of the number of spontaneous nuclear transitions in a time interval dt.

The early unit of activity was the Curie, Ci, defined as the activity of 1g of radium or 3.7×10^{10} disintegrations per second. The curie has been replaced by the SI unit of Bequerel (Bq) where 1Bq is one disintegration per second.

$$1 \text{ Ci} = 3.7 \times 10^{10} \text{ Bq}$$

$$1mCi = 37 \text{ MBq}$$

Activity content can be a useful specification, especially for source transportation regulations, however, it is a difficult quantity to measure and for clinical purposes requires a defined relationship between activity and the radiation emission properties. In order to calculate the output, it was necessary to know the attenuating properties of the source, the encapsulation material and the exposure rate constant. The exposure rate constant, Tδ, is related to activity via the equation.

$$T_\delta = (d^2/A) \, (dX/dt)_\delta \tag{2}$$

Where $(dX/dt)_\delta$ is the exposure rate due to photons of energy greater than δ at a distance d from a point source containing activity A.

13.6.2 Equivalent activity

Due to difficulties in determining the activity content, the equivalent (or apparent activity) was used to specify some early brachytherapy sources. This was calculated from the exposure rate external to the source and the exposure rate constant to obtain the activity of an unfiltered point source that would give the same output.

Care must be taken when using activity to define source strength for clinical dose calculations, as true content activity will yield different dosimetry results than equivalent activity.

13.6.3 Specification by emission

An alternative to using activity is to specify sources in terms of an emission property, and so avoid errors due to uncertainties in exposure rate constant, errors made in allowing for encapsulation will also be reduced.

ICRU introduced the concept of 'kerma' (kinetic energy released per unit mass) in 1962 as an alternative to exposure, to highlight the two stage process in transferring energy to matter from indirectly ionizing particles.

$$K = dE_{tr}/dm \tag{3}$$

i.e. K = sum of initial kinetic energies of all charged ionizing particles liberated by uncharged ionizing particles in mass dm.

A BIR/IPSM report in 1993 recommended the use of 'reference air kerma rate' (RAKR) which is the kerma rate to air, in vacuo, at a reference point which is 1m from the source centre. Recommended units are µGy h^{-1}.

As part of Task Group 43's rethink of the way sources are defined, the AAPM recommends a quantity defined as the product of the air kerma rate at a distance, d, measured along the transverse bisector of the source, and the square of the distance, d. This quantity is the 'Air Kerma Strength' and is assigned the symbol U, where:

$$1U = 1 \, \mu Gy \, m^2h^{-1} = 1 \, cGy \, cm^2h^{-1}$$

Air kerma strength U and reference air kerma rate in µGy h^{-1} although dimensionally different are numerically equal.

New brachytherapy sources received in a radiotherapy department for clinical use are supplied with a manufacturer's certificate that states the source strength, usually in reference air kerma rate. The certificate will indicate the uncertainty in the manufacturer's calibration (eg +/– 5% for a microSelectron source, +/– 7% for stranded I-125 sources) and it is the responsibility of the receiving department to calibrate new sources to independently verify source strength.

It is often not appropriate to carry out measurements at 1m, in vacuo and in scatter free conditions, so recommendations also suggest ionization chamber measurements in air at shorter distances, an appropriate calibration factor is required to convert charge reading to air kerma rate. It also necessary to correct for air attenuation and scattering, for the response of the detector that cannot be regarded as a point detector, and for any deviation from inverse square on extrapolation from actual measurement distance to 1m.

Most new brachytherapy sources are now specified in terms of reference air kerma rate (RAKR), it is this quantity that is verified by measurement by clinical users and subsequently entered into user's treatment planning system. However to achieve clinical patient dosimetry, a calculation must be made within the system to convert the air kerma rate to absorbed dose rate within water.

Traditionally this was calculated using the equation:

$$\text{Dose rate to water} = \text{RAKR.} \; f(r). \; ((\mu_{en}/\rho)_{water}/(\mu_{tr}/\rho)_{air}). \; d^2/r^2 \qquad (4)$$

$f(r)$ = radial function for attenuation and scatter in water

$(\mu_{en}/\rho)_{water}$ = mass energy absorption coefficient for water

$(\mu_{tr}/\rho)_{air})$ = mass energy transfer coefficient for air

d^2/r^2 = inverse square scaling factor.

This has now been superceded using the TG43 U1 formalism where:

$$\text{Dose rate to water} = D(r,\theta) = S_k.\Lambda.[G_L(r,\theta)/G_L(r_0,\theta_0)].g_L(r).F(r,\theta) \qquad (5)$$

S_k = Air kerma strength in U—is the air kerma rate, in vacuo, at a specified distance (usually 1m) along the transverse axis of the source. The unit of S_k is the 'U', where 1U is 1 μGy m^2 h^{-1}. In practice measurements are carried out in air and then corrected for attenuation.

Λ = dose rate constant in water—constant is an absolute value, and is the dose rate to water at 1cm on the transverse axis of a source of 1U. It includes source geometry, spatial distribution of radioactivity, encapsulation and self-filtration within the source and scattering by the medium.

$G_L(r,\theta)/G_L(r_0,\theta_0)$ = The geometry function—accounts for the variation of relative dose due only to the spatial distribution of radioactivity within the source. For distances greater than about two to three times the largest source dimension it will differ from the inverse square law by less than 1%.

$g_L(r)$ = the radial dose function—defines the fall off of dose rate along the transverse axis due to absorption and scattering in the medium.

$F(r,\theta)$ = the anisotropy function—accounts for the anisotropy of the dose distribution around the source, including the effects of self-filtration, oblique filtration of primary photons through the encapsulating material, internal scattering within the source and attenuation and scattering in the surrounding water medium.

See Radiotherapy in Practice:Brachytherapy, Appendix of Chapter 3 for more detail.

Chapter 14

Unsealed sources for therapy

S J Chittenden, G Flux, and B Pratt

14.1 Introduction

Unsealed sources of radioactivity have been used for therapy for over 60 years. They are usually administered in liquid form orally, by intravenous infusion or by direct injection into tumours or body cavities. Capsules containing radioactive iodine are also frequently given orally. The term 'systemic radionuclide therapy' is where the radioactivity is not confined to a tumour or cavity during treatment and implies that the whole body is irradiated. The success of treatment partly depends on how well radioactivity is concentrated and retained in the tumour(s) in preference to the normal tissues. The term 'targeted radionuclide therapy' is when the activity is preferentially accumulated in the tumour(s) either due to direct administration or by selective uptake of the radiopharmaceutical due to biological or chemical processes.

The clinical aspects of radionuclide therapy have been extensively covered in a previous book in this series therefore this chapter will focus on the main physics aspects only.

14.2 Choice of radionuclide

The choice of radionuclide for unsealed source therapy is extremely important and will depend on physical properties such as the type of emissions, range in tissue, the linear energy transfer, the chemical properties of the radionuclide or radiopharmaceutical that are required for localization and retention in the tumour and also on the radiobiological properties such as the half-life and dose rate.

14.2.1 Types of emission

To date the majority of radionuclide treatments have been carried out using beta particle emitters such as ^{131}I and ^{32}P. Radioisotopes such as ^{123}I that produce Auger electrons (see Chapters 2 and 3) have also been used and currently alpha emitters such as ^{223}Ra are being explored. Radionuclides that also emit gamma rays of suitable energy for imaging/counting to assess distribution and/or uptake of activity are particularly useful in radionuclide therapy, but the presence of gamma radiation increases the radiation dose delivered to normal tissues. Where pure beta emitters are used for therapy there is the potential to image/count the bremsstrahlung radiation produced.

14.2.2 Range in tissue

It is important to match the range of the particles to the size of the lesions being treated. If the particle range is high and the lesion size small then much of the particle energy may be deposited outside the lesion causing reduced absorbed dose to the lesion and unnecessary dose to normal tissue. The range of a particle depends on its energy and there is a wide range of beta emitters now available for radionuclide therapy with a variety of energies and applications.

Auger electrons have a very short range (generally <1 μm) and are therefore only of use when attached or very close to the cell nucleus. Alpha particles have a longer range (typically 50–90 μm), which is equivalent to several cell diameters. For both of these particles, heterogeneity of uptake is a particular problem in radionuclide therapy.

14.2.3 Linear energy transfer and specific ionization

A charged particle travelling through a medium loses energy and produces ionization as it travels. The linear energy transfer (LET) of particles has been discussed in Chapter 3 and Chapter 4. Particles with high values of LET are important for radionuclide therapy as they cause more biological damage to a cell than those with low values. Values for alpha particles are typically 100 times greater than for electrons of the same energy.

14.2.4 Chemical properties

The chemical properties of a radiopharmaceutical determine its behaviour in the body. Radiopharmaceuticals are designed for maximum uptake in the tumours or normal organs of interest and minimum uptake elsewhere in the body so they need to remain chemically stable for as long as possible. The shelf life of a radiopharmaceutical (generally quoted as time after activity reference date) is stated by the manufacturer. Examples for therapy radiopharmaceuticals are 2–6 weeks for $Na^{131}I$ capsules, 4 weeks for $^{89}SrCl$ liquid and 2 days for $m^{131}IBG$. There may be other stipulations for product preparation for example $m^{131}IBG$ delivered frozen must be used within 2 h of defrosting and dilution. Radiopharmaceuticals prepared in-house have no specific shelf life (e.g. many antibody and peptide preparations) therefore the final product must undergo stringent quality control testing before use. If the shelf life of a product is exceeded then the radiopharmaceutical will become increasingly chemically unstable. This will result in non-specific uptake of radioactivity in the body and unwanted radiation dose in these areas. It is important to note that even where shelf life is relatively long, the radionuclide will decay physically during this time and the activity remaining may be too low for effective clinical use. Drug interactions must also be considered. Chemical interactions may affect the behaviour of the radiopharmaceutical therefore it may be necessary to alter the patient's medication prior to administration.

14.2.5 Half-life

The physical half-life of a radionuclide is the time taken for the activity to decay to half of its original value. The biological half-life of a radionuclide in a tumour or organ of the body is the time taken for the biological retention of the radioactivity to reduce to half of its original value and does not include physical decay. When direct measurements are taken of radioactivity in the body the measurements include the effects of both physical and biological decay and the effective half-life (T_e) is related to the physical (T_p) and biological (T_b) half-lives as shown in Equation 1.

$$\frac{1}{T_e} = \frac{1}{T_p} + \frac{1}{T_b}$$

(1)

14.2.6 Dose rate

In contrast with external beam radiotherapy, radionuclide therapy delivers continuous irradiation at low and decreasing dose rates. A given dose has a reduced biological effect if delivered at low dose rate as the irradiated cells have more time to repair damage between ionizations; however early responding tissues such as the bone marrow are less affected by this reduction than late responding tissues such as the kidney. This results in sparing of late responding normal tissues in the later stages of therapy hence the differential sparing between tumour and normal tissues is maximized. The dose rate in a particular tumour or organ during radionuclide therapy varies according to the effective half-life of the radioactivity in that volume. The linear-quadratic model for cell survival as used in external beam radiotherapy may be adapted for radionuclide applications and may be useful when choosing a radionuclide for therapy.

14.3 Dosimetry techniques

European regulations state that all radiation exposures are individually planned and doses to normal tissues kept as low as reasonable practicable (ALARP). In the past most radionuclide therapy prescriptions were for standard amounts of radioactivity and absorbed doses to tumours and normal organs were rarely calculated. However there is now evidence of a dose responce (rather than an activity responce) relationship in radionuclide therapy. This means that patient specific dosimetry should be carried out in radionuclide therapy to both optimize the treatment for each patient and to compare doses between groups of patients to improve treatment.

For individual patient dosimetry estimates there are a number of possibilities. Firstly (and ideally) a small 'tracer' amount of the therapy radiopharmaceutical may be administered to the patient and dosimetry measurements carried out as described in Sections 14.3 and 14.4. The doses delivered to tumours and normal organs of interest are then calculated and the amount of radioactivity to be administered for therapy is determined. The tracer and therapy doses should occur reasonably close in time, however for I-131 NaI treatment of thyroid disease there is a possibility of reduced therapy uptake due to 'stunning' if they are too close together. The second option is for the 'tracer' dose to be a different radiopharmaceutical more suitable for dosimetry measurements, eg ^{123}I mIBG or ^{124}I mIBG before to therapy with m^{131}IBG, and ^{111}In Dotatate before ^{90}Y Dotatate therapy. Another alternative occurs in patients who attend for multiple treatments within a few months of each other where there are no substantial changes in the patient's condition between therapies. Here the activity for the first therapy might be decided empirically or from a weight based formula. Absorbed doses are calculated from dosimetry measurements during therapy. These are then used to calculate the activity required for the second therapy. Measurements made during therapy 2 are used to calculate administered activity for therapy 3 and so on.

The most widely used method of calculating absorbed doses for patients treated with unsealed sources is known as the Medical Internal Radiation Dose (MIRD) system. Originally developed for diagnostic nuclear medicine it has now been adopted for radionuclide therapy.

14.3.1 MIRD theory

In the MIRD system, the body is considered to be composed of source organs and target organs. Source organs are those that have taken up radioactivity and target organs are those for which the absorbed dose must be calculated. Frequently the source and target organs are the same. If we consider the simple case of a single source and single target the first step is to calculate the amount of radioactivity taken up into the source organ and the length of time it remains there. The number of radioactive disintegrations occurring in the source organ in this period is known as the cumulated activity $(\tilde{A})$ and is equal to the time integral of the activity in the source organ:

$$\tilde{A}_s = \int_0^\infty A_s(t)\mathrm{d}t \tag{2}$$

which is the area under the activity time curve $A_s(t)$ for that organ.

The next step is to calculate how much energy is emitted from each radioactive disintegration occurring in the source organ and what fraction of it the target absorbs. These factors will be different for each separate type of emission produced in the disintegration. The mean energy emitted per nuclear disintegration for radiation of type i (alpha, beta or gamma) is termed the equilibrium absorbed dose constant (Δ_i) and depends on the energy and frequency of the emission. The fraction of the radiation of type i emitted from the source and absorbed by the target is called the absorbed fraction (ϕ_i) and depends on the type and energy of the emission, the geometry of the source and target, and the composition of the tissues between them. For non-penetrating radiation ϕ_i is 1 when the source and target organs are the same, 0.5 at a source target interface and 0 when source and target organs are physically separated.

Once the parameters described above have been determined the following equation is used to calculate the absorbed dose $D_{t\leftarrow s}$ (in Gy) to the target organ from the activity in the source organ:

$$D_{t\leftarrow s} = \frac{\tilde{A}_s}{m_t} \sum_i \Delta_i \phi_i \tag{3}$$

where $\tilde{A}_s$ (in MBq.h), Δ_i (in g Gy MBq^{-1} h^{-1}) and ϕ_i are as defined above and m_t is the mass of the target organ in g.

The mean dose per unit cumulated activity $\frac{1}{m_t} \sum_i \Delta_i \phi_i$ is known as the S value and equation 3 simplifies to:

$$D_{t\leftarrow s} = \tilde{A}_s S_{t\leftarrow s} \tag{4}$$

In general there will multiple source organs in the body and the total dose to each target organ is simply the sum of the contributions to each target from the individual source organs.

14.3.2 **MIRD calculations in practice**

14.3.2.1 Determination of cumulated activity, $\tilde{A}_s$

In order to calculate the cumulated activity in a source organ it is necessary to determine the activity in the organ at multiple time points following administration of the radionuclide. In practice there are a number of ways of measuring the activity and this is discussed further in Section 14.4. These data are used to generate an activity time curve and in order to describe the curve accurately it is vital to ensure that the number and frequency of the activity measurements are adequate. The cumulated activity is equal to the area under the curve. In the case of a single organ or tumour the time activity curve is usually described by a single exponential equation:

$$A_s(t) = A_0 \exp(-0.693t/T_e) \tag{5}$$

where A_0 is the initial activity in the source organ (assuming rapid initial uptake), and T_e is the effective half-life of the radiopharmaceutical in that organ. In this case the cumulated activity $\tilde{A}_s$ is given by:

$$\tilde{A}_s = 1.44 A_0 T_e \tag{6}$$

In some cases however (e.g. when considering the whole body as the source organ) the activity time curve is generally more complex and is described by the combination of more than one exponential. An example of a single phase dose calculation for a thyroid remnant from ablation with $Na^{131}I$ is shown in Table 14.1.

Table 14.1 Calculation of dose to thyroid remnant during ablation therapy with 3.7 GBq

$Na^{131}I$

Data used:

$\overline{E}_\beta$ = 0.192 MeV

Δ_β = 0.099 g Gy MBq^{-1} h^{-1}

φ_β = 1

T_e = 193 h (physical half-life of ^{131}I)

A_0 = 92.5 MBq (assuming 2.5% uptake)

m_t = 8 g

Using eqns (3) and (6), the dose to the thyroid remnant from activity in the remnant is given by:

$$D_{remnant \leftarrow remnant} = \frac{1.44 \times 92.5 \times 193}{8}(0.099 \times 1) \text{ Gy}$$

$$= 3213 \times 0.099 \text{ Gy}$$

$$= 318 \text{ Gy}$$

NB For simplicity only the principle beta radiation has been used in this example.

14.3.3 **Determination of *S* values (and mass of target organ, *m*ₜ)**

S values have been tabulated for a variety of radionuclides and for different source target combinations in configurations representing a standard man, and more recently a pregnant woman and children of various ages, as well as various other phantoms. Corrections may be applied for organs of non-standard mass however to do this it is essential to determine the mass as accurately as possible. This may be achieved through high resolution imaging such as ultrasound, CT and MRI, however it should be noted that the radioactivity may not be distributed through the entire anatomical volume of the organ. Imaging with SPECT or PET may be used to estimate functional volume, but will have relatively poor spatial resolution. *S* values do not exist for tumours therefore tumour doses must be calculated.

14.3.4 **Limitations of the MIRD scheme**

MIRD has its limitations. The assumption is made that the radioactivity is uniformly distributed in the source organ which is unlikely to be the case in practice. A further assumption is that the organ sizes and shapes in the patient are the same as those in the standard human phantoms used in the scheme. Also the dose calculated is only a mean absorbed dose in the target organ and the maximum and minimum doses may differ significantly from this. Nevertheless the MIRD system itself is elegant and accurate and is very useful for radionuclide therapy dose calculations. Application of the MIRD schema on a voxel by voxel scale can obviate the assumptions regarding mean distributions of activity and mean absorbed doses.

14.3.5 **Olinda**

A useful and easily available computer program to calculate mean absorbed doses to organs and tumours is the Olinda software. This is based on the original MIRD system, but includes more recent organ models, additional radionuclides and dose models for alpha emitters. The software can also calculate dose estimates for spheres of different masses, used for tumour absorbed dose estimates.

14.3.6 **Limiting organs**

In radionuclide therapy, the absorbed doses delivered to the normal organs will limit the maximum dose that can be delivered to the tumour. This will depend on the radiopharmaceutical, the routes of administration and excretion. For example in systemic radionuclide therapy the dose limiting organ is usually the bone marrow, but for intrathecal administrations it will be the spinal cord. Doses to normal organs can be reduced, e.g. by giving amino acids and diuretics to reduce kidney dose in ^{90}Y Dotatate therapy, or by hydration to reduce bladder dose. A bone marrow harvest may be taken before therapy for re-grafting after therapy if necessary. When new therapy agents are introduced, extensive dosimetry and toxicity studies are essential to determine the dose limiting organs.

14.4 **Counting and imaging for activity estimation**

Activity quantification may be carried out on a variety of different counting and imaging systems from simple Geiger counters to complex PET cameras. The choice of system to use will depend on the radionuclide and count rate to be measured and also on other factors such as equipment availability and acceptable measurement time. Regardless of the system used there will be aspects of the counting procedure that introduce uncertainty in the final activity estimation. It is therefore necessary to avoid, reduce or correct for these effects during the quantification process. The most important issues to be considered in counting and imaging for dosimetry are outlined briefly below.

14.4.1 **Counting time**

The counting or imaging period must be long enough so that adequate counts are acquired, but not too long so as to be uncomfortable for the patient. This is because the percentage error in the recorded count reduces with increasing counts.

14.4.2 **Dead time**

All counting systems have an inherent 'dead time' which is the length of time a system takes to process a count after detection and during which no other counts can be processed. When the count rate incident on a detector is high (such as in the early days after radioiodine therapy for thyroid cancer) then over the time it takes for a patient measurement, a substantial proportion of the true counts may be lost. It is essential that a correction is made to account for this loss during activity quantification. For each detector system there will be a maximum incident count rate above which it is not possible to correct adequately and the only alternative is to reduce the incident count rate by, e.g., the use of low sensivity collimators on a gamma camera.

14.4.3 **Activity calibration**

Detector systems register counts. These must be converted into corresponding activity levels in order to calculate cumulated activity. This conversion is achieved using a calibration factor. An example of a simple calibration method is generally used in whole body measurements with Geiger counters. The patient is counted immediately after radiopharmaceutical administration and before their first void. As no activity has been excreted the calibration factor is simply the recorded count divided by the administered activity. Subsequent patient counts are multiplied by this factor to convert them to activity levels. In the case of SPECT calibration factors one option is to scan a phantom representative of the patient's body that contains one or more radioactive sources of known activity and volume representing the tumours (or normal organs) of interest. The scan is carried out with identical scan parameters to the patient scan. After appropriate corrections have been made to the images (as outlined elsewhere in this section) the counts in the patient tumour and phantom source are found. The tumour counts are converted into activity by multiplying by the calibration factor which is given by the phantom counts divided by the phantom activity.

14.4.4 **SPECT vs planar imaging**

When imaging on a gamma camera a choice must be made between planar and SPECT imaging. Both methods have been used for activity quantification in the past although there has been little published data comparing their use in dosimetry. A SPECT scan will have poorer resolution than the equivalent planar scans, but has the major advantage of providing 3-dimensional (3-d) data with improved contrast. 3-d data is generally essential in dosimetry to determine the precise pattern of activity uptake and assess changes in distribution over time. SPECT/CT and PET/CT scans are highly useful in this respect. Although the use of planar scans for dosimetry purposes is not ideal, there are some occasions when it is necessary for example where the count rates are too low for SPECT to be used.

14.4.5 **Attenuation and scatter**

Radiation passing through a medium will be attenuated and scattered by the medium. These effects have been described in Chapter 3 of this book. Attenuation and scatter not only result in degradation of image quality but can lead to large errors in quantification. Therefore, corrections are made for these effects when carrying out dosimetry calculations for radionuclide therapy. Correction methods vary from the very simple (e.g. using scatter windows during gamma camera imaging to estimate the proportion of scatter in the image) to the highly complex (e.g. the use of a registered CT scan to generate an attenuation correction map for an image).

14.4.6 **Registration**

In order to determine the pattern of activity uptake in the body over time as accurately as possible and to generate the optimal attenuation correction map for a patient it is necessary to register sets of scans to each other. On PET/CT and SPECT/CT scanners the PET or SPECT image will be automatically registered to the corresponding CT (however it is important to note that the images are not acquired simultaneously therefore there is still the possibility of patient movement between the scans). In other situations reasonable registration of scans may be achieved by using fiducial markers placed at specific locations on the patient's skin for each scan. Registration is then achieved by overlaying the scans and aligning the markers. Although registration errors will generally be greater with this method than with the use of PET/CT or SPECT/CT scans with care the errors may be minimized. Body motions such as breathing will also affect the accuracy with which any two scans may be registered.

14.4.7 **Partial volume effect**

The partial volume effect describes the fact that when a radioactive point source is imaged it will appear larger than its true size. Corrections for partial volume effect must be applied where possible otherwise substantial errors in image quantification may result, particularly in estimates of activity concentration.

Chapter 15

Radiation protection

C Taylor, M Waller and P Bownes

15.1 Introduction

Radiation causes illness, injury and death at high doses, and it is assumed there is some risk even at very low doses.

This chapter will concentrate mainly on protecting those working with radiation and those receiving incidental exposures. They receive no benefit from their radiation exposure, which must therefore be restricted to keep their risk at an acceptable level.

15.2 Effects of radiation

The earlier chapters will have given you some idea of the effects of radiation. We can divide them into two types: those that are certain to happen if the dose of radiation is large enough—we call these tissue reactions—and those that might happen at any dose—stochastic effects.

15.2.1 Tissue reactions—deterministic effects

With large doses of radiation, tissue reactions occur. Unless the effect is deliberately brought about for therapeutic purposes, the tissue reactions will be harmful. The International Commission on Radiological Protection (ICRP) calls these effects 'harmful tissue reactions', preferring this term to the previously used 'deterministic effects'. The therapeutic effects of radiation, and the side effects seen in radiotherapy patients, are examples of tissue reactions. There is no justification for anybody other than a patient to receive a dose that would cause a tissue reaction.

Radiation has to damage enough cells to produce a noticeable effect. This requires a relatively large dose of radiation known as a **threshold dose**. If the radiation dose is increased above the threshold dose, more cells are damaged and the effect becomes more severe. Any tissue in the body can show a reaction to radiation, though the sensitivity of different tissues varies. Once the threshold dose has been exceeded, the range of dose to go from mild to very severe effects is quite small. For example, the threshold dose for skin erythema is about 6 Gy but the dose required to produce severe burns and long term ulceration of the skin is only about 10 Gy.

If the radiation dose is delivered slowly, or in small fractions, tissues have a chance to recover as new cells are produced to replace those destroyed by the radiation. The result is that the threshold dose increases: the threshold for skin erythema is about 30 Gy if the dose is delivered in small fractions, of say 1 Gy per day. Table 15.1 shows the threshold doses for effects on various tissues, for single exposures and fractionated doses.

Table 15.1 Threshold doses for tissue reactions from ICRP 42

Effect	Exposure pattern	Threshold dose
Erythema	Single, brief exposure	6–8 Gy
	Highly fractionated	30 Gy
Epilation	Single, brief exposure	3–5 Gy
	Highly fractionated	50–60 Gy
Haemopoetic syndrome		
Onset of symptoms	Single, brief exposure	1 Gy
LD-50	Single, brief exposure	2.5 Gy–5 Gy
GI syndrome		
Onset of symptoms	Single, brief exposure	3 Gy
LD-50	Single, brief exposure	10 Gy

15.2.2 Stochastic effects

Stochastic is a mathematician's term for 'occurring at random'. Stochastic effects happen when a cell that has been damaged by radiation is imperfectly repaired so that its future behaviour is modified. Such a cell might turn out to be the starting point for a cancer some years in the future or, if it is a cell involved in reproduction, a hereditary effect. In principle, such an effect can be the result of a single interaction in one cell—so there is no threshold dose to speak of. The probability of such an effect will increase as the amount of radiation and the number of cells involved increases. In other words, the probability of a stochastic effect increases with the effective dose. We will learn about effective dose in the next sections.

For people working with radiation or patients undergoing diagnostic tests, doses in normal circumstances are well below any threshold for tissue interactions, so what we are concerned about is the risk of a stochastic effect.

15.2.3 Summary of effects

Harmful tissue reactions, or deterministic effects:

◆ Happen when a given level of dose, known as the threshold dose, is exceeded.

◆ Become more severe with increasing dose.

◆ Become more severe with increased dose rate.

Stochastic effects:

◆ Do not require a threshold dose: they might occur at any radiation dose, however small.

◆ Are more likely to happen as dose increases.

◆ Are more likely to happen if the dose is received over a short period of time.

◆ Do not become more severe as the dose increases.

◆ May occur many years after the radiation exposure.

15.3 **Dose quantities and units**

In Chapter 7 we learnt about absorbed dose, measured in Gray (Gy). Absorbed dose is appropriate for prescribing tumour doses because it relates to tissue reactions. However, it does not relate directly to the probability of a stochastic effect. At the low doses encountered in work with radiation we need a quantity that will give at least an indication of this probability: in other words, a measure of risk. To do this we need to take into account not only the dose, but also what type of radiation gave rise to the dose, which tissues have been irradiated and how sensitive those tissues are to radiation. **Effective dose** is a quantity designed to do this.

15.3.1 **Equivalent dose and effective dose**

Some types of radiation seem to be riskier than others for the same absorbed dose: one milligray received from exposure to alpha particles is more likely to result in a cancer than the same absorbed dose from gamma rays. To take this into account, each type of radiation has been assigned a weighting factor. Multiplying the absorbed dose (Gy) by the radiation weighting factor gives a quantity called 'equivalent dose', measured in sievert (Sv). For photons (X-rays and gamma rays) the factor is 1, so that the equivalent dose is numerically equal to the absorbed dose. For other types of radiation, the weighting factors are greater (see Table 15.2). In effect, the weighting factor indicates how much more damaging the type of radiation is than X-rays. This can be written as an equation, where H is the equivalent dose, D is the absorbed dose and W_R is the radiation weighting factor.

$$H = D.W_R$$

If there is more than one type of radiation making up the exposure, the equivalent doses from each type are added together.

The weighting factor is loosely based on a quantity called the 'relative biological effectiveness' (RBE) of the radiation. However it is not equal to the RBE. Weighting factors have been rounded to values such as 1, 10 and 20. For most of the radiation you are likely to encounter in radiotherapy work, the radiation weighting factor will be 1, so that the mean absorbed dose to a tissue, and the equivalent dose will be numerically equal.

The distribution of the dose within the body is taken into account by using tissue weighting factors. For any tissue, the equivalent dose multiplied by the tissue weighting factor gives the effective dose, also expressed in sievert. Again, this can be written as an equation, where E is the effective dose and W_T is the tissue weighting factor.

Table 15.2 Radiation weighting factors

Radiation type	Weighting factor, W_R
Photons: X and gamma rays	1
Electrons	1
Neutrons	5 to 20, depending on energy
Alpha particles	20

$$E = H.W_T = D.W_R.W_T$$

The factor for each tissue is based on an estimate of the risk posed by the dose to the tissue, as a fraction of the risk if the whole body received the same dose. Since irradiating the whole body is worse than irradiating part of it, the tissue weighting factors are all less than 1. Usually when a person is irradiated, more than one tissue receives some dose. In this case the effective dose is worked out by adding together the contributions to effective dose from all the tissues.

$$E = H_1.W_{T1} + H_2.W_{T2} + \ldots$$

The tissue weighting factors were revised in 2008, and published by the ICRP. The values are shown in Table 15.3. Note that:

- Not all tissues are listed,
- There are only a few different values for weighting factors, several tissues having the same weighting factor.

This does not mean that the unlisted tissues are somehow immune to radiation or that, e.g., the lung is exactly as radiosensitive as the stomach. It means that calculating the effective dose in this way gives a sufficiently good estimate of risk to reflect our present knowledge. For any practical situation, the estimate of the risk to a person exposed to radiation would not be improved by refining the weighting factors further.

Table 15.3 Tissue weighting factors

Tissue	Weighting factor, W_T
Gonads	0.08
Bone marrow (red)	0.12
Breast	0.12
Colon	0.12
Lung	0.12
Stomach	0.12
Bladder	0.04
Liver	0.04
Oesophagus	0.04
Thyroid	0.04
Bone surface	0.01
Brain	0.01
Salivary glands	0.01
Skin	0.01
Remainder	0.12
Total	**1**

You can see that calculating effective dose results in a single number, whatever the pattern of radiation exposure. This allows comparison of radiation exposures from quite different types of exposure. For example, a frequently used comparison is that a chest X-ray gives about the same radiation dose as a return flight between the UK and Spain. Whether this reassures patients having X-rays or frightens holiday makers is another question. A chest X-ray involves exposure of just part of the body to X-rays. A flight involves exposure of the whole body to enhanced levels of cosmic radiation, which is a mixture of gamma rays and various exotic types of particle. Nevertheless, the process of calculating effective dose results in a similar number for each exposure.

15.3.2 Background radiation

Although we may think of radiation as something artificial that comes from X-ray sets or nuclear power stations, there are many natural sources of radiation. For most people natural sources are the greatest source of radiation dose. In the UK, the Health Protection Agency reviews the different contributions to natural background radiation and periodically publishes its findings most recently in 2005.

Natural sources of radiation include cosmic radiation, gamma radiation from the surroundings—mainly building materials, radioactivity in the body—mainly from food, and radon gas. On average, we all receive about 2.2mSv per year from background sources with most of it coming from radon. In areas such as the south west of England, where radon concentrations are higher, the average dose is more like 6mSv per year.

15.4 Risks

Effective dose gives us a single number that is intended to indicate the risk of an effect such as cancer induction. What we mean by risk is the probability that some adverse effect will occur. At present, the accepted estimate of the fatal cancer risk from radiation is 1 in 20 000 per 1 mSv of effective dose.

15.4.1 Linear no threshold model

Our estimates of radiation risk are based on studies of people who have received relatively large doses, often over a very short time—for example the Japanese survivors of the atom bombs dropped at the end of the Second World War. The pattern of exposure is quite different from what we expect in occupational exposure to radiation, where the doses are usually smaller and are received over many years rather than a few hours or days.

In applying the high dose and high dose rate data to low doses and dose rates, a 'linear no threshold' (LNT) model has been used. This assumes that the risks for small doses are in proportion to those for the larger doses in the studies, and that there is no threshold dose. While there is good reason to suppose that the processes that produce effects at high dose will also occur at low doses, it is quite possible to use other models, some of which are illustrated in the graph. Any of these models can be made to fit the epidemiological data, and each will produce a quite different estimate of the risk at low doses. The ICRP has stated that the LNT model has been retained 'for simplicity'.

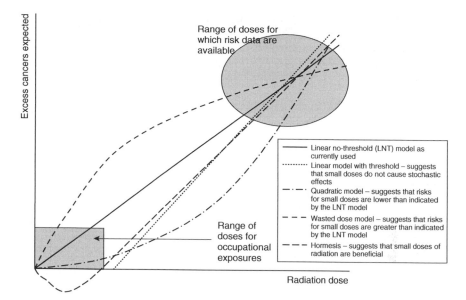

Fig. 15.1 Graph illlustrating the linear no threshold model.

15.4.2 Getting from risks to dose limits

For tissue reactions, setting dose limits is a reasonably simple process. You know how much dose it takes to produce a harmful effect so you can choose a level well below this as a limit. Anybody who keeps to the limit is safe. This is what is done when setting limits for the exposure of hands, feet, eyes and small areas of skin.

With stochastic effects, legislators have more of a problem. Any limit greater than zero implies some risk, and people often oppose the imposition of risks. For this reason the ICRP has put forward principles for radiation protection:

♦ Exposure to radiation must produce a positive net benefit—in other words, it must do more good than harm.

♦ Exposure to radiation must be kept as low as reasonably achievable (ALARA), taking into account economic and social factors.

♦ Exposure to radiation must be kept below the limits.

15.5 Legislation

In the UK, there are several pieces of legislation that relate specifically to radiation. The main ones that you need to know about are:

♦ The Ionizing Radiations Regulations 1999 (IRR99);

♦ The Ionizing Radiation (Medical Exposure) Regulations 1999 (IRMER);

♦ The Environmental Permitting (England and Wales) Regulations 2010 (EPR10) and the Radioactive Substances Act 1993 (RSA);

♦ The Medicines (Administration of Radioactive Substances Regulations 1977 (MARS),

15.5.1 **The Ionizing Radiations Regulations (1999) (IRR99)**

These regulations concern health and safety in the workplace, so they are aimed at protecting people who are working with radiation and anybody else who happens to be present. They are not aimed specifically at hospital work. The regulations are enforced by the Health and Safety Executive (HSE), which has powers of inspection.

The regulations are supported by an 'approved code of practice' and by official guidance from the HSE. In addition to these two documents, guidance relating to hospital work has also been published. The regulations themselves say very little about practical measures, leaving this to the various guidance documents. It is up to the employer to put in place protection measures and then to demonstrate, to the satisfaction of an inspector, that these measures are sufficient.

A summary of the requirements of the regulations is given in Table 15.4.

Most of the requirements apply to the employer. Employees must avoid deliberate exposure to radiation, and comply with reasonable requirements that the employer puts in place to satisfy the regulations.

The employer must take 'all necessary steps to restrict so far as is reasonably practicable' exposure to radiation. The phrase 'so far as is reasonably practicable' is very important. It means that to comply with the regulations it is not sufficient merely to keep exposures below the dose limits; doses must be reduced further so long as this remains reasonably practicable. On the other hand it is not necessary to take inordinate measures to reduce doses that are already low. In order to decide what is necessary, the employer has to do a risk assessment, and keep a record of it. The regulations specify three levels of intervention for restricting exposure:

◆ Use engineering controls (prevent people from irradiating themselves);

◆ Use systems of work (tell people not to irradiate themselves);

◆ Use personal protective equipment (accept that some exposure to radiation is inevitable, and do what you can to reduce it).

The regulations incorporate dose limits based on those recommended by the ICRP. There are limits on equivalent dose to individual tissues, aimed at ensuring that tissue reactions are avoided altogether, and there are limits on effective dose aimed at reducing the risk of stochastic effects to 'acceptable levels'. These are shown in table 15.5.

Note that these limits do not apply to patients. The exposure of patients is covered by the ionising radiation medical exposure regulations (IRMER). However, when a patient is not actually undergoing a medical radiation exposure, he or she is 'any other person'. Note also that an employer will usually treat an employee who does not work with radiation as 'any other person'.

15.5.1.1 **Controlled areas**

Areas where people might receive high doses have to be separated from other areas by designating them as controlled areas. The regulations define controlled areas as somewhere where a person might exceed 6mSv per year. They go further than this and say that if people can avoid high doses only by following special procedures, you still need to control the area.

Once an employer has decided to control an area they have to demarcate it. This means letting people know that the area exists by putting up warning signs. He then

Table 15.4 Summary of the ionizing radiations regulations (1999)

Regulation	Topic	Requirement
7	Prior risk assessment etc.	Decide on the precautions to be put in place
8	Restriction of exposure	Make sure doses are ALARP—use engineering controls, systems of work and personal protective equipment. Restrict the exposure of pregnant women. Investigate high doses.
9	Personal protective equipment.	Provide it and test it.
10	Maintenance and examination of engineering controls etc. and personal protective equipment.	Decide on a test procedure Do the tests. Record the results.
11	Dose limitation.	Ensure doses do not exceed limits.
12	Contingency plans.	Have plans to reduce the effect of any accident that the risk assessment shows to be probable.
13	RPA	Appoint and consult a suitable RPA
14	Information, instruction and training.	Inform employees of risks, precautions and requirements of regulations. Give others enough information to keep them safe.
15	Cooperation between employers	If more than one employer has employees at a place of work, the employers must cooperate to ensure they all comply with the regulations.
16	Controlled and supervized areas.	Identify areas where precautions are required to prevent high doses. Restrict access to such areas.
17	Local rules and radiation protection supervisors.	Write local procedures that will ensure compliance with the regulations. Appoint people to supervize staff so that they follow the local rules.
20–25	Classified persons and dose assessment.	Identify employees who are likely to exceed 6 mSv per year. Arrange medical surveillance and personal monitoring for them.
27–30	Arrangements for keeping radioactive substances.	Keep radioactive substances securely. Ensure all sources are accounted for.
31,32	Radiation generating equipment, including equipment for medical exposures.	Manufacturers must design equipment properly. Equipment must be installed correctly and only used for appropriate purposes. Equipment must be tested at appropriate intervals. Quality assurance programmes must be devised and followed.
33	Misuse of radiation sources.	Employees must not willfully misuse radiation sources.
34	Duties of employees.	Employees must not deliberately expose themselves to radiation, and must report faults.

Table 15.5 Dose limits

Limit on equivalent dose to hands, feet and any 1cm² of skin	
Category of person	Annual dose limit (mSv)
Employee	500
Trainee under 18 years old	150
Any other person	50
Limit on equivalent dose to the lens of the eye	
Category of person	Annual dose limit (mSv)
Employee	150
Trainee under 18 years old	50
Any other person	15
Limit on effective dose	
Category of person	Annual dose limit (mSv)
Employee	20
Trainee under 18 years old	6
Any other person	1

has to keep out unauthorized people. The only people allowed to enter controlled areas are 'classified persons' and people who enter under a 'written arrangement' designed to keep their dose below 6mSv per year. So, are you authorized to enter a controlled area? Read your local rules to find out!

15.5.1.2 Classified persons

If an employer thinks that his employees are likely to exceed 6mSv per year, he has to classify them. Classification brings about a new set of requirements for the employer to comply with: annual medical reviews to ensure that employees are fit to be classified, routine measurement of individual doses and keeping individual dose records for 50 years. Understandably most employers would prefer to avoid this, and this means that they have to ensure that doses exceeding 6mSv are 'unlikely'.

Most NHS staff are not classified because, with good procedures, it is relatively straightforward to ensure they do not exceed 6mSv per year. The main reason that most NHS staff have personal dose monitoring is so that the employer can demonstrate that doses are much lower than 6mSv. Records of dose measurements done under these circumstances only need to be kept for 2 years rather than 50.

15.5.1.3 Radiation protection adviser

The two preceding sections show that the employer has a few decisions to make. Are staff going to exceed 6mSv per year? Do they need to set up controlled areas? Which staff need personal dose monitoring?

To get answers to these and other questions the employer has to consult a Radiation Protection Adviser (RPA). The RPA is somebody, usually a medical physicist, with expertise in radiation protection demonstrated by having a 'Certificate of Competence'. Appointment of an RPA is a requirement of the regulations.

15.5.2 **Ionizing Radiation (Medical Exposure) Regulations (2000) (IRMER)**

Radiation doses received by patients for diagnosis or treatment are regulated by this piece of legislation. The general aims of IRMER are:

◆ To ensure that medical exposures are used only when they are likely to be beneficial,

◆ To ensure that appropriate levels of dose are used,

◆ To ensure that those responsible for irradiating patients are appropriately trained.

At the time of writing, the regulations are enforced by the Care Quality Commission (CQC). In the past they were enforced by the Department of Health and then by the Healthcare Commission, and it is possible that future reorganizations may result in further changes.

As with IRR99, the employer carries general responsibility for complying with the regulations but certain individuals also have responsibilities:

◆ **Referrer**—the person who wants a patient to have a medical exposure;

◆ **Practitioner**—the person responsible for justifying a medical exposure;

◆ **Operator**—a person carrying out practical aspects of a medical exposure.

A typical situation envisaged by the legislators involves a doctor (the referrer) who wants to have his patient X-rayed. He sends a request to the X-ray department where somebody (the practitioner) vets it to ensure that the X-ray is appropriate for the clinical indications. The patient then sees a radiographer (the operator) who actually does the X-ray.

In radiotherapy the situation is different. The doctor who sends a patient to a clinical/radiation oncologist is not requesting radiotherapy even if they suspect that this would be the most appropriate treatment, so he is not a referrer. If the oncologist decides that radiotherapy is appropriate and requests it for his patient, he is acting as both referrer and practitioner.

For any individual treatment there will be a number of operators. They will include those who plan and simulate the treatment, those who deliver the treatment and perhaps those involved in the calibration of the equipment.

The regulations do not determine who acts as referrers, practitioners and operators. Referrers and practitioners must be 'registered healthcare professionals' but they, and operators, must also be 'entitled by the employer's written procedures to act in that capacity'. So it is up to the employer to say who is allowed to have responsibility under the regulations. While it might be expected that practitioners are more likely to be radiologists and operators are more likely to be radiographers, it is the employer's procedures that actually determine the roles. The regulations do not suggest that any role should be filled by any particular profession.

The employer's procedures must specify who the referrers, practitioners and operators are, and the people chosen must know when they are acting in this capacity. One of the most common criticisms of radiotherapy departments inspected under the regulations is that their procedures do not make clear the responsibilities of individuals.

The most important requirements of the regulations are:

◆ To have written procedures,

◆ To allocate responsibilities,

◆ To ensure that all exposures to radiation have been justified by a practitioner,

◆ To ensure that the practitioner and operator cooperate to ensure that all radiation exposures of patients are optimized,

◆ To ensure that practitioners and operators are adequately trained.

15.5.2.1 Justification and authorization of exposures

The regulations distinguish between justifying radiation exposures and authorizing them. Justification is the decision that a radiation exposure is appropriate, and this decision must be made by a practitioner. Authorization is the process of demonstrating that justification has taken place. This can be a signature on a treatment request or some other form of recording. A practitioner can authorize an exposure personally, or can issue written guidelines that allow an operator to authorize it. Potential confusion as to whether somebody is justifying an exposure or authorizing it under guidelines further illustrates why clear written procedures are essential.

15.5.2.2 Medical physics experts (MPE)

The employer must employ graduate scientists 'experienced in the application of physics to the diagnostic and therapeutic uses of ionizing radiation'. MPEs must be involved in all therapy exposures, and are normally central to the treatment planning process and the development of quality assurance procedures. They need not be directly involved in every diagnostic exposure, but they must be available for consultation.

15.5.2.3 Reporting patient overexposures

Both IRR99 and IRMER require the employer to report patient doses that are 'much greater than intended'. IRR99 covers doses that have been caused by equipment faults (you report these to the HSE) and IRMER covers doses caused by anything else (report to the CQC).

At present, the definition of 'much greater than intended' for radiotherapy exposures is the same for both regulations: 10% more than intended for a whole treatment and 20% more than intended for a single fraction. While this definition is at first sight very simple it can be difficult to interpret, for example in the case of a small error in positioning the treatment field.

15.5.3 The Environmental Permitting (England and Wales) Regulations (2010) and the Radioactive Substances Act (1993) (RSA93)

The Environmental Permitting (England and Wales) Regulations 2010 (EPR10) became law in April 2010, at which time large parts of the Radioactive Substance Act (RSA 93) were rescinded. Whereas RSA93 controlled the keeping and disposal of radioactive materials, EPR10 deals with permission to hold and dispose of all manner of potentially polluting substances. The general aims of the two pieces of legislation are similar and, at the time of writing, it appears that the procedures to be adopted when using and disposing of radioactive materials have not changed significantly. In Scotland, the Radioactive Substances Act still applies.

The legislation is intended to protect the general public against hazards from uncontrolled use or disposal of radioactive sources. Before being permitted to use, store or dispose of radioactive materials the user must register with the Environment Agency (EA) or the Scottish Environmental Protection Agency (SEPA). Under EPR10, registrations and authorizations are known as permissions.

Certificates of registration and authorizations for waste disposal are specific to the user and site. You need separate registrations for closed and unclosed materials. The registration certificates specify how many closed sources of each nuclide you can hold, and how active they can be or, for unsealed radioactive substances, they specify what activity of each radionuclide can be held. Standard conditions are issued with the registration covering the supervision, marking, keeping, use and records of the registered materials. It also states what action must be taken if sources are lost or stolen or damaged or unsealed materials are dispersed in an uncontrolled manner. Disposal authorizations specify the activity of each nuclide that can be disposed of by any particular route, for example via the sewage system, by incineration or by disposal to landfill. In order to get a disposal authorization, an assessment of the environmental impact of the disposal has to be submitted.

There are no national waste disposal limits. Each establishment negotiates its own authorizations with the EA on the basis of its environmental assessments. The legislation can have a direct impact on patient care in that an unsealed radioactive substance administered to a patient becomes radioactive waste when the patient goes to the lavatory. The disposal limits set in the authorization can determine the number of patients that can be treated over a given time.

15.5.3.1 High activity sealed sources (HASS) and orphan sources regulations

In 2005 the High Activity Sealed Sources (HASS) and Orphan Sources Regulations were introduced. They define a high activity source in terms of its activity and nuclide. Usually in radiotherapy departments the only sources that come under these regulations are iridium-192 sources used in high dose rate remote afterloading and cobalt-60 sources used in teletherapy and Gamma Knife radiosurgical treatment units. The regulations impose requirements over and above those imposed by RSA93, including:

+ Arrangements for the safe management of high activity sources, including what happens when they are no longer needed. The source owner must guarantee adequate financial provision for the eventual disposal of the source. The aim of this is to avoid situations where the owner no longer uses the source but cannot afford the cost of disposal. The usual approach in UK hospitals is to incorporate a source take back or replacement agreement in the service contract with the supplier.

+ A security plan for the site. Security must be regularly assessed by a counter terrorism security advisor (CTSA). The Environment Agency will ask the local CTSA to assess the security measures of the site when they assess the application to hold the source. The level of security measures is based on the risks posed by the sources being held.

◆ Stringent record keeping for both the users and competent authorities for each high activity source from its manufacture to its disposal.

15.5.4 The Medicines (Administration of Radioactive Substances) Regulations (1978) (MARS)

These regulations apply to any clinician (doctor or dentist) who wants to do something that will make a patient radioactive; for example nuclear medicine tests, radionuclide therapy and brachytherapy. The regulations therefore do not apply to X-ray exposures or external beam therapy.

Anybody wishing to perform such procedures must have a certificate known as an ARSAC licence. ARSAC stands for the Administration of Radioactive Substances Advisory Committee, the body that issues certificates.

To get a certificate, a doctor applies to the committee. The application must detail the qualifications and experience of the applicant, the facilities available for performing the procedures and a statement that there are satisfactory radiation protection measures in place. The application form has to be countersigned by a MPE, the person responsible for providing the radioactive materials (e.g. a radiopharmacist) and a RPA.

There are some things that you need to know about ARSAC licences:

◆ Each licence applies to one person—one doctor cannot take responsibility for the exposure of a patient by working under the licence of another.

◆ Each licence applies to one site—if a doctor does procedures at more than one hospital he will need a licence for each hospital.

◆ Each licence lists the procedures that the holder is allowed to perform—if a procedure is not listed then you cannot do it.

◆ A separate licence is needed for research. You cannot do research exposures under a standard clinical licence.

15.6 Practical radiation protection

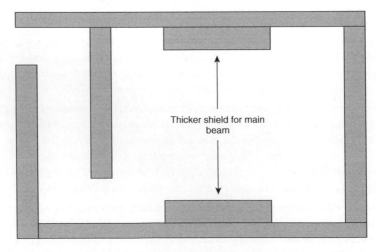

Thicker shield for main beam

Fig. 15.2 Plan of a typical linac chamber with a maze entrance.

15.6.1 **External beam radiation therapy**

In external beam therapy the main protection issue is the extremely high dose rate emitted by the linear accelerator. The dose rate in the primary beam is such that the annual dose limit for an employee would be received in a matter of seconds. Even the scattered radiation within a linear accelerator chamber would cause the limit to be exceeded within a few minutes. Protection therefore involves the use of massive shielding combined with systems to prevent exposures taking place when anybody other than the patient is in the treatment room. This separation of staff from the source of radiation means that doses to staff working in external beam therapy are usually lower than for those involved in brachytherapy or radionuclide therapy.

Not only is the radiation dose rate high but the radiation energy used, with X-rays being generated at anywhere between 6MV and 15MV, means that the radiation is very penetrating. So in a typical linear accelerator chamber the walls that are struck by the primary beam are often around 2.5m thick, if they are made from ordinary concrete. The walls that protect only against scattered radiation, plus radiation that has penetrated the linac housing, will probably be about 1.5m thick.

The way in to the linac chamber is usually through a maze. The maze needs to be several metres long and must incorporate 2 or more changes of direction so that radiation has to scatter several times before it escapes, by which time the dose rate has been sufficiently reduced. If a door were used instead, it would need to weigh several tons to provide the necessary shielding. Consequently, it would need a motor to open and close it. Nevertheless, this is sometimes done if space is at a premium. A compromise is to have a shorter maze with fewer turns and a rather more modest door.

The other major problem is to prevent an exposure from taking place while anybody is inside the room. One measure is to provide an interlock at the entrance. If there is a shield door then opening it will trigger the interlock and switch off the radiation. If there is no door, then a light beam or an interlocked gate across the maze entrance is used. Breaking the light beam will cut off the radiation.

What if somebody is already in the room, and a person outside wants to switch on the radiation? Once the interlock has been triggered by somebody entering (even if the radiation was off at the time), a series of switches has to be pressed by the person leaving, otherwise the radiation cannot be switched on. It would, of course, be possible for somebody other than the last person to leave to do this, so the procedures include a requirement to check that everyone has left before switching on the radiation. A closed circuit TV system allows you to do this.

If, in spite of all this, somebody is still in the room when an exposure is started, there are warning lights and audible warnings, and an emergency cut off switch is provided inside the room.

15.6.2 **Neutrons**

When linear accelerators are operated at potentials greater than 10MV, neutrons are produced. Neutrons and X-rays behave differently, both in terms of how they are transmitted through materials and how they are scattered. Lead is ineffective for shielding neutrons and instead you need materials containing hydrogen. The water content of the concrete walls normally used to shield linear accelerator chambers means that these walls will also be effective at shielding neutrons.

Neutrons tend to scatter along mazes and are not greatly affected by corners. To reduce the neutron dose rate we need to make the maze narrower and longer. The walls of the maze can be coated in neutron absorbing material; materials containing boron are normally chosen.

As a rough guide, the neutron shielding requirements are:

◆ For linacs operating at 8MV or less, neutrons are not produced.

◆ At 10MV, some neutrons are produced, but the concrete walls and maze for protecting against X-rays will be effective for neutrons. Neutrons will be a problem for rooms with no maze, and the design of the door will have to take neutrons into account.

◆ At 15MV, neutron scatter along the maze will be a problem and the maze must be designed to cope with this.

15.7 **Brachytherapy**

This section relates principally to protecting staff involved in brachytherapy procedures or care of brachytherapy patients, and people visiting such patients. You do not have to worry about the extremely high dose rates encountered in external beam therapy, nor about the inevitable contamination that occurs from patients undergoing radionuclide therapy. However, there are protection issues that are unique to brachytherapy.

◆ The sources used give rise to significant dose rates, and the radiation energy is usually high enough for any shielding to have to be thick.

◆ Some contamination can occur when sources such as ^{192}Ir are used in the form of wire that has to be cut to the required length.

◆ The physical size of the sources is small enough for them to be lost or stolen, so security is an issue.

Examples of current brachytherapy techniques are:

◆ Manually loaded brachytherapy—permanent iodine-125 seed prostate brachytherapy.

◆ Manual afterloading—temporary iridium-192 wire implants.

◆ Remote afterloading—high dose rate iridium-192 stepping source systems.

Three important radiation protection measures are used:

◆ Minimize the **time** spent exposed to radiation, as the total dose received is directly proportional to time.

◆ Maximize the **distance** from the source of radiation, as dose received follows the inverse square law.

◆ Use **shielding.** This is done by placing a radiation absorbent barrier, often lead, between the source and the area where people need to be.

15.7.1 **Principles of afterloading brachytherapy**

Compared with manual implantation of sources, afterloading systems in brachytherapy reduce radiation doses to staff and are better for optimizing the dose distribution to the patient. Afterloading refers to any method where applicators are placed in tissues or body cavities so that sources can be loaded later. This allows more time and care to be taken in positioning the applicators without the risk of receiving any radiation exposure.

Afterloading of sources can be done either by hand, or remotely by specially designed equipment that drives the sources into position. Although **manual afterloading** eliminates the radiation dose to the clinician in theatre who positions the applicators, the person who inserts and removes the sources is still exposed to radiation. Treatments usually last for a few days, during which time the patient needs normal medical care, so nursing staff also receive some radiation exposure.

In **remote afterloading** systems the radioactive sources are contained in a lead safe close to the treatment couch. The system is connected to the applicators that have already been inserted into the patient, and the sources are transferred into the required position either by a pneumatic transfer system (LDR/MDR ^{137}Cs units) or via drive cables (^{192}Ir HDR units). The sources can be retracted from the patient and returned to the safe if any nursing is needed during the treatment, so the treatment room can be treated in much the same way as an external beam therapy room, with shielding and interlocks. Therefore no radiation exposure to staff should occur unless the equipment malfunctions or the interlocks fail.

15.7.2 Controlled areas for brachytherapy

Most areas where brachytherapy sources are present are designated as **controlled areas.** The specially designed rooms for remote afterloading units and dedicated sealed source storage rooms will be permanently controlled. Areas such as theatres, recovery rooms and wards where sources are handled, or where there are patients with sources inside them, will be controlled for as long as the sources are present. The status of such areas (controlled or not) has to be shown on a warning notice.

For the duration of their treatment, patients with implants are accommodated in specially designed rooms, with lead shielding to reduce the dose rates outside. Inside the room moveable lead shields are used to reduce the dose to people who need to enter the room for caring or visiting. The dose rates near the patient are measured so that safe working practices can be devised and restrictions on visiting times can be calculated. Nursing and theatre staff must be properly trained. Theatre and ward staff will normally require personal monitoring badges.

15.7.3 Quality assurance issues relating to brachytherapy radiation protection

Source Storage:

- A centralized source inventory is required to demonstrate compliance with the registration limits under RSA 93 or EPR10.
- Radioactive source stores should be shielded and secure. Regular audits should be performed to account for all the sources.
- All sources should be identified individually.
- The location of every source should be known at all times.

- Brachytherapy sources must be tested for leakage annually or when damage of the source is suspected.
- Any radioactive sources which have reached the end of their recommended working life should be taken out of clinical use and disposed of.

Receipt of sources:

- The strength of all sources should be independently measured by the user and compared to the calibration certificate before clinical use. Measurements should be within ±5%.
- Autoradiography is a simple method used to check the **distribution of activity** within a source, source train configuration or source position.
- Sealed radioactive sources should have been leakage tested by the manufacturer. If this has not been done the user should test them before clinical use.

15.7.4 Source preparation (not including remote afterloading units)

Radioactive sources should always be handled using long handled forceps to reduce doses to the fingers. More substantial shielding, with holes to put your arms through, is used to reduce doses to the body and transparent shields are used to minimize dose to the eyes. It is important for operators to be able to work quickly to reduce the time they are exposed to radiation, so training and practice of techniques is essential. When sources are not in use they are stored in shielded containers. Cutting iridium wires to obtain the correct length and activity produces contamination in the form of small wire fragments, so all handling and cutting tools and the preparation area have to be monitored at the end of source preparation.

15.7.5 Source movement within the hospital

Sources should only be moved by trained staff. Movement has to be done as quickly as practicable, without leaving the sources unattended, and without allowing other people to come close to the source.

15.7.6 Manual brachytherapy treatment delivery and nursing care

Table 15.6 summarizes some general points that should be part of systems of work for theatres and wards.

15.7.6.1 Patient related issues

15.7.6.2 Temporary Implants

Temporary implants are designed to be removed at the end of the treatment. During the treatment the patient stays in hospital, so imposing controls is relatively easy.

- Before patients can be discharged from the hospital all the radioactive sources must have been removed and accounted for.

Table 15.6 Work procedures for manual brachytherapy

Area	System of work procedure
Everywhere	◆ Clear warning signs. ◆ All sources should be accounted for at all times. ◆ Monitor equipment and area after use. ◆ Use special tools to maximize distance when handling sources. ◆ Clear lines of communication between staff.
Theatre	◆ Personal dose monitors and lead aprons should be used when appropriate. ◆ Only send for sources when required. ◆ Place sources behind protective barrier when not used. ◆ Check source information is correct.
Wards	◆ Personal dose monitors should be worn. ◆ When removing sources, check the number and position of sources against information provided. ◆ Define time limits and distance restrictions where practicable for staff and visitors. ◆ All operations should be carried out in the minimum time consistent with accuracy and safety. ◆ Maximize distance and use shielding, when appropriate. ◆ Have a portable radiation monitor available. ◆ Monitor everything leaving the room to ensure no sources present. For permanent prostate implants monitor the urine for sources. ◆ Shielded container available for a patient with removable sources.

◆ If the patient needs surgery the temporary implants are removed from the patient before surgery commences.

◆ If a patient dies when a temporary implant is still *in situ* then the radioactive sources and applicators are removed immediately and the body is monitored before it can be released for a post mortem examination or disposal.

15.7.6.3 Permanent implants

These are usually seeds (small sources less than 1mm long) of either iodine-125 or palladium-103, often used to treat the prostate. The sources are relatively short lived and are not meant to be removed. So for a few months after the implant some restrictions are applied. These are listed on an instruction card given to the patient when he goes home.

◆ Information and an instruction card should be given to the patient about potential hazards and restrictions of contact between the patient and others.

◆ A patient receiving permanent iodine-125 or palladium-103 seed implants of the prostate poses a relatively low risk and can be discharged on the day of the implant. He is advised to avoid prolonged periods of very close contact with children and pregnant women for the first two months post implant.

◆ If the patient needs surgery, the RPA should be consulted. The RPA will provide advice on the precautions needed or on whether elective surgery needs to be delayed.

- If a patient dies after a permanent implant then there are not normally any special restrictions on burial. Radiation risk assessments and environmental impact assessment should be made to determine if cremation is permitted.

15.7.6.4 Contingency procedures in brachytherapy

Because we are using radioactive sources, we have to think about the things that might go wrong, and plan for what to do when it happens. The plans we make are called contingency plans or contingency procedures. These have to be written down, and staff have to be trained so that they know the procedures and how to implement them. Some of the things we have to plan for are:

- **A lost or stolen source.** The RPA and RPS must be informed immediately. The RPA will normally coordinate the search for the source and the investigation of the incident. Unless the source is found straight away both the HSE and the EA have to be informed. Losing a source is embarrassing enough. Losing one and not realizing it for several days is even worse, and is likely to lead to prosecution of the hospital. This is one reason why maintaining accurate source records and doing source audits is so important.

- **A damaged sealed source.** If it appears that a source might have been damaged, a report has to go to the person responsible for it—usually the brachytherapy physicist. The aim is then to test the source to see if it is leaking, and to minimize the spread of contamination. If it is leaking, it needs to be put into an airtight, shielded container. Areas where the source has been and people in those areas need to be monitored to see if they are contaminated. It might be necessary to inform the EA and the HSE, and the RPA is the person to advise you about this. Disposal of the source can then be arranged.

- **Fire.** The Fire Service will want to know what sources there are and where they all are, in case they need to attend a fire. You need to have a centralized sealed source inventory indicating the type, number of radioactive sources and their location. This information along with a layout of the premises should be given to the local fire service.

- **Source retraction failure.** Emergency procedures for remote afterloading units must cover the possibility of a source retraction failure. The risk of this occurring can be minimized by quality assurance of the unit and the applicators used. Part of the daily QA procedure should ensure that all equipment required for the emergency procedure is available.

15.7.6.5 Brachytherapy room design

The design of a room will depend on its purpose. A room that houses a remote after-loading machine will contain a high activity source all the time. It will require automatic systems to prevent entry to the room while the source is exposed, together with automatic monitoring and warning systems. Such rooms are not used for any other purpose.

On the other hand ward side rooms in which patients with implant sources are treated will not have any automatic systems. Sometimes they are used for non-radioactive patients, or for radionuclide therapy patients, so clear warning signs are essential to let staff know what type of hazard is present. Both types of room need shielding appropriate to the type of source in use.

The RPA has to be consulted at the design stage about all the radiation protection aspects, particularly the shielding, and he will normally liaise with those responsible for designing and constructing the room to ensure that these aspects are considered at every stage.

The design should ensure that dose constraints are not exceeded. A dose constraint is a level, below the dose limit, that takes into account the possibility that people might encounter radiation from more than one source. A typically chosen dose constraint is 0.3mSv per year for people other than radiation workers. Whenever possible the dose rate outside the room should be reduced to a level at which it can be classed as a non-designated public area. The treatment room itself will be a controlled area.

The shielding requirement will depend on: the proposed treatment techniques and the typical dose rates of these treatments, the number of patients to be treated, duration of the treatment and whether the room has other uses needing special requirements. The radionuclides to be used and their maximum air kerma strength will influence the shielding requirements. Table 15.7 shows some of the properties of the common nuclides used in brachytherapy. You can see from the half and tenth value thicknesses that quite thick shielding is needed to achieve even modest dose rate reduction.

The **location of the brachytherapy room** is very important, as it needs to be close to the sealed source room, theatres and imaging facilities (X-ray, CT, MRI).

Radiation doses to surrounding areas have to be estimated and confirmed through radiation monitoring. Occupancy of these areas will influence the dose estimates.

Table 15.7 Basic properties of common brachytherapy radioactive sources

Nuclide	Principal γ Energy (MeV)	Half-life	Typical reference air kerma rate(μGy h^{-1})	Transmission through lead (mm) HVT	TVT	Transmission through concrete (mm) TVT
^{60}Co	1.17, 1.33	5.27 years	HDR remote afterloader up to 5.7 x 10^3	12	40	206
^{137}Cs	0.66	30.17 years	LDR remote afterloader 28 to 115	6.5	21	157
^{192}Ir	0.3–0.6	73.83 days	Hairpin/wire 0.13–1.3 per mm HDR remote afterloader up to 42 x 10^3	4.5	15	147
^{125}I	0.035	59.4 days	Seeds for permanent prostate implant 0.1 to 1 per seed	0.03	0.1	–

HVT half value thickness.
TVT tenth value thickness.

Table 15.8 Occupancy factors for various areas

	BIR/IPEM (11)	NCRP 49 (12)
X-ray room, offices, staff room	100%	1
Corridors	20%	¼
Toilets	10%	1/16
Stairways, waiting rooms	5%	1/16

Occupancy factors have been published by different bodies, and some of these are shown in Table 15.8. Brachytherapy sources are not collimated so the orientation factor in all calculations will be 1.

Other requirements to be considered are:

◆ Doors must be shielded. It is not always practical to put enough shielding into a door, so it may be necessary to have additional shields in the room so that the sources do not irradiate the door directly.

◆ A portable radiation monitor should be available near to, but outside the treatment area.

◆ Radiation warning notice at the entrances to the room.

◆ CCTV and intercom system.

◆ Space requirements, including access for bed trolleys.

◆ En-suite toilet facilities.

◆ For remote afterloading systems additional requirements are required:

• For HDR units a small maze maybe necessary to ensure no direct irradiation of the lead lined door and to reduce scatter at the door.

• Door interlocks.

• In addition to inbuilt system monitors there should be an independent radiation monitor that indicates when the source is out.

• An illuminated warning sign to indicate when the source is in the treatment position.

• Clearly labelled emergency stop buttons.

• Emergency mechanisms for source retraction failure or power failure.

◆ If a diagnostic X-ray unit is to be used within the treatment room to obtain localization radiographs this will need its own warning system and protection measures.

The IPSM report 75 shows examples of typical treatment rooms for: a manual Cs-137 afterloading room for gynaecological intracavitary treatments, low dose rate/medium dose rate remote afterloading and high dose rate remote afterloading facilities.

15.8 **Radionuclide therapy**

Therapy with radionuclides involves the use of relatively high activities of unsealed radionuclides. Preparation and administration of the radioactive material presents a number of radiation protection problems in terms of exposure to high dose rates and the possibility of contamination. Administration of the material results in patients who are radioactive. Depending on the nuclide used, they may emit significant dose rates (over $250\mu Svh^{-1}$ at 1 m) and they are also a potential source of high levels of contamination. The fact that the source of radiation and contamination is a human being with all the physical and social needs that that entails makes control of these hazards more difficult. The radiation protection arrangements have to take all these problems into account.

15.8.1 Individual risk assessment

The radiation protection arrangements, based on an individual risk assessment of every patient, must identify adequate solutions to all these problems before the final decision is taken to proceed to treatment. In particular the prior risk assessment must take into account the radiation consequences of the care that a patient may require post-treatment. This must include the extent to which patients can care for themselves without close proximity help, whether they are continent of urine, and whether they can understand and will comply with precautions to restrict the exposure of others. Foreseeable contingencies (e.g. cardiac arrest or acute surgical intervention) that would necessitate close contact with the patient must be planned for. Patients' child care responsibilities must be considered in detail.

The practitioner for the radionuclide treatment is responsible for an individual administration and must hold an ARSAC licence (discussed above). At the initial consultation patients should be provided with information detailing treatment administration, effectiveness, potential risks and radiation protection restrictions in place during and after the treatment. The patient should not have any contraindications to treatment and they must sign a consent to treatment form indicating that they understand the radiation protection advice and (where appropriate) that they are neither pregnant nor breast feeding.

15.9 Radiopharmacy

The radiation protection measures required for the preparation of therapeutic and diagnostic radiopharmaceuticals are not qualitatively different. However, the shielding requirements for aseptic isolators may be greater for therapy work, and the consequences of a spill of a long lived radionuclide may mean that provision of a separate isolator is advisable. As with brachytherapy treatments, a safe means of transporting therapeutic doses to the point of use must be devised.

Table 15.9 lists some common radionuclide therapies with the principal radiation hazard and references to detailed procedure guidelines.

15.9.1 Pure beta emitters

Pure beta particle emitters, such as ^{90}Y, ^{89}Sr and ^{32}P pose comparatively small risks to others, provided the risk of contamination is small. There will be a small external dose rate, due to bremsstrahlung, but this is unlikely to be significant unless long periods of close contact with the patient are unavoidable. Treatment with pure beta emitters can usually be done as a day case procedure provided the probability of a contamination incident is assessed to be low.

Table 15.9 Common radionuclide therapy procedures

Radionuclide/ pharmaceutical	Use	Admin route	Primary excretion route	Hazard Internal	External	Guidelines/ procedures
I-131 NaI	Benign thyroid disease	oral	renal	yes	yes	RCP 2007 (29), IAEA safety series no. 63 (30).
I-131 NaI	Thyroid cancer	oral	renal	yes	yes	RCP 2002 (31), IAEA safety series no. 63 (30). Eanm.org/ scientific_info/ guidelines
I-131 mIBG	Neuroectodermal tumours	Slow I.V	renal	yes	yes	IAEA safety series no. 63 (30). Eanm.org/ scientific_info/ guidelines
I-131 Lipiodol	Hepatocellular carcinoma	Selective arterial injection	Renal	yes	yes	IAEA safety series no. 63 (30). Eanm.org/ scientific_info/ guidelines
I-131 antibody	Lymphoma	IV	Renal— slow	yes	yes	IAEA safety series no. 63 (30).
Sm-153 EDTMP	Palliation of bone pain	IV	renal	yes	yes	IAEA safety series no. 63 (30). Eanm.org/ scientific_info/ guidelines
Re-186 HEDP	Palliation of bone pain	IV	renal	yes	yes	IAEA safety series no. 63 (30). Eanm.org/ scientific_info/ guidelines

Table 15.9 (continued)

Radionuclide/ pharmaceutical	Use	Admin route	Primary excretion route	Hazard		Guidelines/ procedures
				Internal	External	
Sr-89 SrCl	Palliation of bone pain	IV	Renal	yes	no	IAEA safety series no. 63 (30). Eanm.org/ scientific_info/ guidelines
Ra-223	Palliation of bone pain	IV	Renal/ faecal	yes	no	
Y-90 peptide	Neuroendocrine tumours	IV	renal	yes	no	IAEA safety series no. 63 (30).
Lu-166 peptide	Neuroendocrine tumours	IV	renal	yes	yes	IAEA safety series no. 63 (30).
Y-90 Sirspheres	Hepatic tumours	Selective arterial injection	none	no	no	IAEA safety series no. 63 (30).
Y-90 colloid	Installation into cysts	injection	none	no	no	
P-32 phosphate	Myelo- proliferative disease	IV	renal	yes	no	IAEA safety series no. 63 (30). Eanm.org/ scientific_info/ guidelines
Ho-166 DOTMP	Multiple myeloma	injection	renal	yes	no	IAEA safety series no. 63 (30).
Various	Radiation synovectomy	Injection	None	No	No	IAEA safety series no. 63 (30). Eanm.org/ scientific_info/ guidelines

15.9.2 Shielded side rooms

If the patient is not continent, or treatment is undertaken with gigabequerel (GBq) amounts of a gamma emitter such ^{131}I, then an inpatient stay is likely to be essential to restrict the exposure of others. The patient will stay in a shielded side room with en-suite bathroom and toilet with floors and surfaces that are non-absorbent and easy to decontaminate.

They will stay until most of the urinary excretion has occurred and the external dose rate has reduced to a level that permits the patient to be discharged safely. As an example, for thyroid ablation with 3–5 GBq of ^{131}I, this is likely to take three to four days.

The main factors in restricting exposure for an inpatient are as follows:

- If the patient is catheterized, the catheter bag should be placed in a lead pot.

- Patients should be as self-caring as possible to keep staff doses low. Access to the treatment room is restricted to trained personnel who must adhere to good working practices that incorporate the three principle concepts of radiation protection: time, distance and shielding.

- All staff that enter the patient's room or handle potentially contaminated items must wear plastic overshoes, aprons and gloves. These are discarded in designated plastic bags on exit from the room and treated as radioactive waste.

- Staff must monitor themselves for contamination immediately after removal of protective clothing. If contaminated, staff should proceed according to defined procedures.

- Items should not be removed from the patient's room unless first checked for contamination by trained staff.

Routine sampling of body fluids should be avoided in the first few days of therapy when these samples are likely to be most radioactive.

A key element of achieving good radiation protection is to ensure that patients understand that they will effectively be in isolation for several days and to ensure they have their preferred books, DVDs, computer games etc. available to occupy them during their stay. During the inpatient stay, a dose constraint will be set for visitors, which means that visiting time will be limited and visits by children and pregnant women may have to be deferred.

The side room will be a controlled area, but an area just inside the room is usually marked off for visitors. Staff entering further into the room ('crossing the line'), must wear gloves and overshoes, which are removed upon leaving the room—as will be described in the system of work written in the local rules. A contamination monitor must be available at all times. Nursing and domestic staff must be trained to follow a system of work that restricts exposure. When it is known in advance that a patient will require close nursing care post-treatment, it is important to use personal integrating dosemeters to monitor the doses received.

15.9.3 Discharges to sewer

The route taken by the waste pipe from the en-suite toilet to the point of discharge to a main sewer must be as direct as possible so that, in the event of a blockage, other lavatories not intended for radioactivity are not affected. The waste pipe may need to be shielded if it passes close to occupied areas. It must be labelled to indicate the presence of radioactivity. Unless unusually high activities are being discharged, hold up tanks (to allow waste to decay before being discharged) should not be necessary.

However, when designing new facilities, consideration should be given to how hold up tanks could be installed, if the restrictions on discharges were to be tightened.

15.9.4 Laundry, crockery and solid waste

In addition to the shielded room itself, facilities are also required for dealing with contaminated linen, contaminated crockery and cutlery and (generally low level) radioactive clinical waste. While most waste can be stored and later disposed of as very low level waste, long half-life material such as ^{89}Sr may have to be consigned to a contractor for disposal while still radioactive.

15.9.4.1 Comforters and carers

Comforters and carers are defined as 'individuals who (other than as part of their occupation) knowingly and willingly incur an exposure to ionizing radiation in support and comfort of another person who is undergoing, or who has undergone, a medical exposure' by the UK HSE. A comforter and carer must be informed of the likely exposure they may receive and of the associated risks. They must also receive instruction and information on how to keep their dose ALARP. Dose limits for comforters and carers do not apply but it is good practice to restrict the dose less than 5 mSv from one series or course of treatment.

15.9.4.2 Patient guidance following release

Following radionuclide administration, patients should be given an information card giving the details of the administration, the duration for which special precautions are necessary and contact telephone numbers. The card should be carried by the patient during the time these precautions are necessary. Radiation protection advice should take into account the patient's personal or social circumstances. This is done with guidance from the RPA or MPE and may be a generic radiation risk assessment, using assumptions for a typical patient. However, if assessment of a patient shows that they do not conform to the generic assumptions, an individual risk assessment is necessary. Typical restrictions include:

- The patient should avoid prolonged close contact with children and pregnant women for a period of time related to the activity of radionuclide retained. If the patient has young children it may be advisable for them to be accommodated elsewhere during this period.
- The patient should avoid prolonged physical contact with family members.
- The patient may need to take time off work for a period related to the activity of radionuclide retained and the nature of their work.
- There may be restrictions on travel related to the activity of radioiodine received. If possible the patient should avoid using public transport for a few days.
- Patients should be informed that some screening procedures at ports and airports may detect residual radioactivity for several months particularly after radioiodine administration.

- Pre-menopausal women should be advised to avoid pregnancy for a few months following treatment.
- Advice from the RPA or MPE should be sought as soon as possible if patients are to be discharged to another hospital or nursing home or to their own home where they receive support from a district nurse.

Table 15.10 shows an example of the instruction card detailing radiation protection restrictions following administration of 400 MBq ^{131}I for thyrotoxicosis (guidance 25).

Table 15.10 Instruction card detailing radiation protection restrictions following 400 MBq radioiodine for thyrotoxicosis

Patient Name	Hospital no.
Address	Consultant
Isotope I-131	Activity 400 MBq

Date of administration:

Guidance for patients treated with radionuclides

Much of the radioactivity will be excreted from your body in urine within a day or two following the administration. The remainder may remain for several weeks. This in turn means that you can irradiate other people who are physically close to you.

The following simple precautions will allow the treatment to be given without causing harm to your family, friends and other persons and without undue restrictions on your daily living.

Avoid close contact with other people (Try to keep at least 1 metre away)	for 2 days
Sleep separately from your partner (Beds should be at least 2 metres apart)	for 3 days
Avoid close contact with children aged 5–16 years (Try to keep hugging and holding to a minimum)	for 11 days
Avoid close contact with children aged 3–5 years (Try to keep hugging and holding to a minimum)	for 16 days
Infants (3 years and under) should be cared for by someone else	for 21 days
Avoid close contact with pregnant women (Try to keep at least 2 metres distance)	for 21 days
You should not go to work	for 2 days
You should not go to social events, cinema, restaurants etc	for 2 days
Avoid long journeys (2 hours or more) on public transport	for 5 days
Avoid becoming pregnant or fathering a child	for at least 6 months

Other

I understand the restrictions explained to me by:	Date:
Signature of Patient or representative	Date:

15.9.4.3 Contingency plans

Contingency plans must be in place for any foreseeable adverse incident. The most likely adverse incident to occur is a spill of body fluids. Written procedures for dealing with spills to hard surfaces and for personnel decontamination should be readily available. Procedures should be in place for other possible events e.g. deterioration in the patient's clinical condition requiring urgent diagnostic investigation or transfer to a critical care unit. In the event of death of a patient soon after receiving unsealed source therapy, the RPA should be notified immediately. Precautions may be required (e.g. for post-mortem and/or regarding disposal of the corpse by burial or cremation).

Chapter 16

Quality assurance in radiotherapy

T Jordan and E Aird

The only real mistake is the one from which we learn
nothing.
John Powell, composer

16.1 Introduction

A radical radiotherapy dose of 70Gy is over 3,500 times the 20mSv annual limit of
exposure for a worker, and around 10 to 15 times the whole body fatal dose that might
be received, from a nuclear accident or weapon detonation. The twin factors of fractionation and restriction of exposure by collimation and targeting in radiotherapy
permit effective use of this lethal potential, and as accuracy in delivery also increases
are now allowing study of the effectiveness of even higher 'dose escalation' regimes.
Nevertheless the potential for damage to the patient is high and as techniques have
become more complex the scope for a procedural or systematic error have increased.
As the delivery has become less transparent to the user, control has increasingly come
to rely on computer management and a procedural system of control and checking.

Possibly every professional involved in radiotherapy delivery will at some point in
their career have reason to face their own fallibility; but by affecting the fundamental
process the major accidents are generally caused (or at least not prevented) by physicists.

> In 1988 a physicist at the Royal Devon & Exeter Hospital made a simple mistake in the
> calibration of a new cobalt source. When the dosemeter over ranged on 1 minute read-
> ings he proceeded to measure for 0.8 minute. In the subsequent calculation the correc-
> tion for this was forgotten and the result was a 25% overdose of 207 patients before the
> error was discovered.

16.2 Is it QA or is it QC?

The Royal Devon & Exeter Enquiry heavily influenced the whole ethos of today's practice.
The main changes consisted of:

1. The concept of *definitive calibration*, which formalized that for all critical measure-
    ments where there is no existing baseline (e.g. a source change, installation of a new
    linear accelerator ion chamber or calibration of a new dose meter), the measure-
    ments would be performed completely independently by another qualified person.

Box 16.1 Quality Assurance and Quality Control

The **Quality Assurance** standard is the managed process to ensure a system of work designed to achieve a specific level of consistency in delivery of a service. It includes clear definition of responsibilities, documented procedures, control of records, documentation and analysis of failures, audit of process and continual improvement. **Quality Control,** on the other hand, is the specific tests or monitoring to ensure adequate standards are maintained, such as making measurements on the linear accelerator beam performance. Quality control tests by definition are a relatively simple and quick to perform subset of measurements designed to adequately check that the detailed measurements at commissioning are maintained.

2. A measure that all centres should participate in independent dosimetry audit with a collection of adjacent radiotherapy centres in an audit group under the umbrella of the Institute of Physics and Engineering in Medicine (IPEM).

3. The production of, the Bleehen Report (WHO, 1988) which introduced quality assurance. This was based on the industrial standards of ISO9000 designed to deliver high quality production processes. Although at first odd to consider the patient as a product this was quickly adapted to health care and processes became formally documented and unsigned and undated memos and datasheets were swept away.

In 1991 a physicist at the North Staffordshire Hospital introduced a new planning computer. During commissioning checks it was confirmed that the previous unit, that had been operating since 1982, had been calculating the monitor units for isocentric treatments correctly (a minority of treatments at that time). However a pre-existing procedure of the planning team, manually correcting for distance using an inverse square law correction, had persisted. This meant 1045 patients were under dosed to various degrees depending on deviation from 100cm FSD. The enquiry looked at managerial lines of responsibility and invested responsibility in the Principal Physicist to institute such a programme or programmes of tests and checks, recurrent or otherwise, that each Clinical Oncologist in the Department is continually assured that any dose of radiation which is prescribed is delivered to the tumour in precisely the manner and the intensity prescribed.

16.3 The radiotherapy prescription

It may seem odd that when the ethos is of second (or even third) checking everything, the prescribing clinician finds they bear sole responsibility that the individual patient treatment is appropriate; but there are a number of safeguards which can assist them in this.

- ◆ Each centre is required by the Network Cancer Standards to have standard protocols for treatment sites. This includes: the clinical indication, the modality, the tumour localization and planning technique, dose/fractionation and overall time, the treatment and immobilization technique. It is an IR(ME)R requirement (see Chapter 15) that the practitioner has all necessary diagnostic information (history, radiology, pathology, surgical report) to justify the exposure. It is quite possible to prescribe non-protocol dose levels but it is good practice to annotate the justification for this (most often patient condition), perhaps as part of the quality system.

◆ There should be a protocol for each treatment site for geometric verification of the treatment that specifies at least the following (cancer standards):

- The imaging equipment and methods used;
- The treatment techniques
- The frequency and timing of imaging (for verification purposes)
- Recommended anatomical reference points
- Tolerance and action levels, which requires a good understanding of systematic and random errors.

◆ When writing a dose prescription (and with reference to the standard protocol) the clinician should define the total dose, the number of fractions and also the dose per fraction. This reduces the chance of errors, particularly in handwritten prescriptions.

◆ Participation in radiotherapy trials usually involves a comprehensive QA check of the treatment process. Target outlining exercises and feedback are particularly useful for the clinician. The trials unit operate graphics software which can display and overlay the contours submitted by all participants allowing direct visual comparison (http://rttrialsqa.dnsalias.org/).

◆ The planning team generally have considerable experience in the typical definition of target volumes and of the practice of different oncologists within a centre. This not only provides a powerful quality check for the genuine error, particularly in a protocol driven system, but in a healthy situation utilizing that expertise in case review or department education can reduce insularity in practice (see also QA in clinical trials).

16.4 The checking process

All steps of the treatment plan go through a series of checks and safety procedures to ensure integrity of the data.

◆ Every single parameter of the treatment plan verified and signed by the operator is separately checked. These include: patient data and prescription, correct CT slice, treatment centre move from reference tattoos (visible on CT as radio-opaque markers), that dose volume histograms of both the target volume and organs at risk meet constraints, that all machine parameters (gantry and collimator angles, field size, wedges, MLC etc.) are correct, that the dose distribution is optimized and meets requirements and that associated verification images (see below) are correct. For a simple (palliative) treatment which does not require computer planning a similar process is followed but a third operator may be involved in manual checking.

◆ The monitor units (MU) calculated by the computer are independently verified. This can be performed with printed data books from commissioning allowing a manual calculation, or with the same data assembled in a spreadsheet. MU calculation is described in Chapter 10. The accuracy of this calculation is typically aimed at around 3%. Remember the planning computer uses quite sophisticated algorithms and corrects for inhomogeneities using CT electron densities and for non-uniform tissue surface variations such as in breast irradiation. The manual check is limited but good agreement can generally be achieved if lung and bone depths are corrected to an equivalent path length in unit density material (see Chapter 10). There are also a number of commercial MU calculation software programmes available that, once commissioned with basic beam data, can provide accurate independent calculation including MLC and IMRT fields.

> *In 2006, a 15 year old girl in Glasgow was given a 56% overdose to the brain. This was an uncommon treatment for medulloblastoma. To increase efficiency, it used to be common practice to calculate the MU for a nominal 1Gray per fraction and then multiplied by the correct dose per faction once the prescription was finalized. However, the centre had changed over to entering the correct MU to start with. The main criticisms from the enquiry report were the out of date procedure in the QA system and lack of an independent MU check.*

16.5 Verification (see also 'reporting of errors' below)

Admit your errors before someone else exaggerates them.
Andrew V. Mason

All radiotherapy departments incorporate an IT network which links CT scanners, simulators, planning computers and treatment machines. Once checked, the treatment is approved and available on the network. The linear accelerators (linacs) use these data for daily fractions and record all exposures given. The machine parameters, MU, mode and energy of treatment, MLC positions and motorized or dynamic wedge data are downloaded and can be auto set without having to transcribe or re-enter data (see also Chapter 5).

> *In 2004, a patient in Leeds received a serious overdose of 135% when the data loaded manually into the linac control system failed to include the planned wedge. The delivered MU were large to account for the wedge attenuation. Subsequently, direct electronic transfer of data was a recommendation published in 2008 in 'Towards Safer Radiotherapy', to eliminate such human transcription errors in the future.*

Moreover, all data for machine and couch angles and positions have attached tolerances appropriate to the technique; the linac control computer will prevent beam initiation on the accelerator until any discrepancies are resolved. At initial treatment the patient is carefully set up using the machine crosswire and distance meter relative to tattoos, together with lateral laser crosses on tattoos or indelible ink marks. Having checked all 'set up' details provided from planning or simulator, the couch position is 'acquired' for the computer verification system.

> *Errors in radiotherapy treatments considered to be sufficiently serious to be reported to the inspectorate (Care Quality Commission, CQC) under IR(ME)R are categorized in their annual report. The most common radiotherapy error is a geographic misplacement of the treatment field.*

The couch position can depend on the daily repositioning of the patient so the couch edges are provided with a system of indexed location points. For high precision treatments such as head and neck radiotherapy a baseboard located to the couch provides for precise relocation of immobilization devices bespoke to a single patient. For breast and lung treatment the board located on the treatment couch may allow tilts and variable armrest positions. Although only the couch height, longitudinal, lateral displacements and rotations are computer verified, every movement on the board has a form of labelled indexing and these are recorded on the patients' set up notes. Even for patients with minimal relocation a simple head rest or knee wedge (e.g. to reduce hip rotation in prostate radiotherapy) can all be fixed to position. These bring day to day variability of the patient on the couch, and in turn the couch position readouts, to within surprisingly narrow tolerances.

The combination of appropriate immobilization fixtures and computer tolerance on reproducibility of daily relocation of the patient (from couch position readouts) can help to greatly reduce errors such as misidentification of a tattoo or a wrong directional move from there.

16.5.1 Image verification

During CT based treatment planning digitally reconstructed radiographs (DRRs, see Chapter 10, Section 10.4.1) are created and attached to the plan. Alternatively the patient may have their treatment set up on the simulator and a genuine digital radiograph taken from an image intensifier. We have discussed above that the patient is initially set up relative to marks using the crosswire and lasers, and then the acquired couch parameters can assist in eliminating any subsequent gross errors. But what if an error was made at this point and the wrong couch parameters stored? In recent years almost all new treatment machines have included an electronic portal imaging detector (EPID) which unfolds behind the patient and gives a direct image of the treatment field as the X-rays exit the patient. This image can be compared to the DRR to assess geometric accuracy (see Chapter 6, section 6.5).

> EPIDs are made of amorphous (i.e. non-crystalline) silicon which can be formed in thin sheets on plastic, similar to LCD TV displays. By doping and etching, the silicon can be made into a 2D array of hundreds of ionization sensitive transistors.

There are three potential problems with this:

- Firstly it is often difficult to locate position from a small treatment portal and so a larger treatment portal is exposed to include surrounding (bony) anatomy;

- Secondly, this procedure gives extra dose to those tissues (1–2cGy per image);

- But thirdly it is a MV image, which is not ideal. The physics of X-ray attenuation is such that at kV energies a small proportion of interactions will be by photoelectric effect. As this attenuation is proportional to the cube of atomic number it means the combination of bone density and its high atomic number constituents such as calcium and phosphorous cause the skeleton to be vividly imaged and even differences in some soft tissues can be seen. At megavoltage energies the Compton effect dominates and therefore contrast is largely gained from physical density (kg/m^3) (see Chapter 3).

The result is an image of such poor contrast it can be very difficult to make out structures. The image features tend to show only where physical density variations have caused more, or less, attenuation than the surroundings, allowing some visualization of bones and air passages such as the trachea. Accepted practice is to operate an imaging protocol, which will vary from centre to centre but typically would involve day 1 imaging (ideally pre-treatment) to eliminate gross error followed by day 2 & 3 to assess reproducibility and separate out a systematic or repeatable mal-position from the random daily variation in patient position.

16.7 Image guidance

This term can cover a wide variety of modalities from the previously mentioned megavoltage image to use of ultrasound to locate an organ position such as the prostate. The key element to *image guidance* is that the image is taken immediately before treatment, in the treatment position, and this is then used to reposition the patient to align the internal anatomy with the treatment beam. Orthogonal images are required to define the required moves in 3 dimensions. Three of the currently most commonly used methods are:

a) The latest linear accelerators have an additional kV X-ray source mounted on the gantry with the beam orthogonal to the MV treatment beam, and its own image receptor plate (see Chapter 6, Fig 6.4). Set up with highly accurate coincidence at the isocentre, this permits orthogonal MV and kV images. Alternatively, for a small increase in time, the gantry may be rotated 90 degrees to obtain a high contrast kV/kV orthogonal image pair. These mainly image skeletal anatomy.

b) By rotating the kV beam around the patient and using reconstruction software similar to a CT scanner, a cone beam CT (CBCT) image can be created to visualize the soft tissue anatomy. As the whole volume of anatomy is being irradiated simultaneously by the diverging X-ray beam ('cone beam') the scatter contribution degrades the image compared to the accumulated narrow slices of the CT scanner, but still provides detailed images for positioning. The innovative *Tomotherapy* unit, which incorporates a small 6MV linear accelerator in a CT gantry, provides pre-imaging with narrow slice Megavoltage CT. This produces an image with very low scattered radiation but with the loss in contrast inherent in an MV energy beam. The intention of these methods is to image the soft tissue. If the organ itself can be imaged it eliminates the problem of internal movement independent of the skeleton.

c) One way round the problem of organ movement without CT is to implant *fiducials*, generally gold seeds or gold coils (which migrate less). This allows 2D planar imaging of the organ and the skeleton. For example in treatment of pancreatic cancer, fiducial markers can demonstrate inter-fraction movement and intra-fraction movement when combined with the appropriate technology.

By marking key structures the image analysis software can do an automatic match for verification by the operator. As stated earlier, one of the most common errors in radiotherapy is incorrect moves and it requires vigilance to ensure adjustments are not made in the opposite direction to that intended. This software may allow the user to automatically move the patient and treatment couch to correct the error under computer control. Furthermore, if a volume is being matched as from CBCT, the software may also give the option to calculate rotation errors giving 3 linear dimensions (x,y,z) and 3 rotational dimensions (roll, pitch, yaw). It is possible (though not common) to purchase 6-D treatment couches to incorporate all six corrections. In all cases operator verification of the 'matched' volume is essential.

16.8 Adaptive radiotherapy (ART)

If we have a CT image of the patient while on treatment then it opens up the possibility of replanning the treatment on a daily basis, either with or without image guidance. One could then allow for organ movement, change in body shape (e.g. weight loss) or growth or shrinkage of the tumour. However, there are some drawbacks. Apart from the fact that planning and checking of treatments is very time consuming, the on-treatment images are currently less suitable to plan on than the original higher quality CT planning scan. In particular, as the 'cone beam' simultaneously irradiates a large volume of tissue, the Hounsfield numbers which define the attenuation of each pixel and are used in the dose calculation algorithm are much more variable in CBCT due to a volume dependence of scattered radiation changing the response.

16.9 In-vivo dosimetry

Following the Glasgow and Leeds accidents, the Chief Medical Officer published various recommendations on safety in radiotherapy in his 2006 annual report. One of these was the use of *in-vivo dosimetry*.

The use of a radiation dose detector placed directly on the patient can provide a useful end of line check that all processes, calculations and set up, result in the correct dose administration. The advantages are obvious and in some countries an in-vivo dose check is mandatory. However this isn't as easy as it might sound. The aim to infer dose from surface measurements requires the detector to have suitable build up for the photon energy used and factors such as body obliquity and mal-position, particularly in non-uniform fields such as wedged distributions, means false positive indication of errors is not uncommon. To prevent shielding of the tumour by the detector its use is limited to early fractions in a radiotherapy course. The overheads of operating such a system in all patients are significant for a small number of errors detected. Cost benefit discussions on the utility of in-vivo dosimetry are ongoing but the possibility remains that their use will prevent a major accident.

There are two types of detector, both described in Chapter 7:

- The traditional thermoluminescent dosimeter (TLD). The advantages of small size and easy application to the patient (including intra-cavitary) are countered by the required preparation (annealing in an oven), the delay before readout, and an accuracy of around 3% if used with care.

- Electronic detectors such as diodes or MOSFETS (metal oxide semiconductor field effect transistor) detect the ionization produced by X-rays in the electronic component, giving immediate readout. These devices are more accurate and much less labour intensive than TLD but apart from regular calibration, may require correction for field size, energy, obliquity and temperature dependence. The sensitive components are only a few millimetres, but the required build up is often made of steel to minimize overall size.

16.10 Quality control of intensity modulated radiotherapy (IMRT) and arc therapy

IMRT gives specific verification problems due to the changing photon fluences or the complex MLCs movement needed to produce them. The standard plan check confirms only the planned result, not that the complex treatment can be delivered correctly and this fact introduces something quite new—patient specific QC.

In this the patient plan is recreated on a test phantom and the dose recalculated for the phantom shape rather than the patient. Before treatment, dose points may be measured with ion chambers in the phantom or the distribution in a plane checked with either film or an array of ion chambers or diodes. In the latter case software analysis allows comparison of such planar dose distributions with that predicted from the treatment planning system. Because some parts of the distribution may have high gradients a simple percentage dose error is not enough. In a high gradient a large dose error may be resolved by moving only a small distance to get good agreement. This is the basis of the gamma index in which points in the image are checked against tolerances for dose and distance. A common tolerance is 3% or 2mm but this may be site and department specific. If either tolerance is satisfied the point passes the criteria and is added to percentage point pass. By this method one gets an image of the dose distribution clearly showing areas of mismatch. This is an excellent way to make a judgement on accuracy of the planned dose delivery.

> In 2005, a doctor in New York decided to adjust a patient's head and neck IMRT plan to reduce the dose to the teeth. While the patient waited the physicist replanned the treatment but on saving had repeated software crashes. Finally the plan was stored and the patient treated. Four days later a verification plan was produced and run on the treatment machine; at this point it was realized that 3 fractions had been given with the MLC left completely open. The department's QA checking procedure was not followed: the error would have been picked up in normal checking procedures. Furthermore, the lack of MLC movement was not spotted during the delivery of the 3 fractions. The result was that the patient received 39Gy in 3 fractions. The patient became progressively blind and deaf, and eventually died two years later.

16.11 Quality control generally

Extensive comment on quality control of equipment for radiotherapy is beyond the scope of this book; such information is readily available, but to give a flavour of what is required it is worth looking at these requirements for the main piece of equipment: the linear accelerator (linac).

16.11.1 **QC for linacs**

It is the responsibility of the medical physics expert (MPE)(see Chapter 15) to coordinate the QC checks carried out by different groups. The MPE is also required to interpret deviations in terms of the effect that they will have on radiotherapy treatments and advise the clinical oncologists accordingly. As described above, the accuracy of both the geometric and dosimetric parameters must be maintained in order to give optimized treatment, allowing maximum control of the tumour with the lowest morbidity of normal tissues. These parameters are typically maintained to values of 1–2mm and 1%–2%. The QC tests are performed regularly at intervals of time depending on the likelihood of change. Daily checks will include output measurements checks and checks on the set up lasers. Weekly or monthly checks will look in more detail at the various beam parameters such as flatness and energy (of all beams, including all electron energies). New techniques (e.g. IMRT) require new methods for checking. Manufacturers are continually producing new equipment to do these checks. The MPE has to make a judgement as to when it is worth spending money on this type of equipment which may save time (and therefore money!) if it performs correctly and efficiently.

16.12 **Risk management**

This subject includes clinical governance; audit; various matters within legislation and peer review.

16.12.1 **Legislative matters (see also Chapter 15)**

The main regulations involved with external beam radiotherapy are: IRR'99 and IRMER 2000. IRR'99 deals with all regulatory matters concerning the safety of staff and public. IRMER deals with the safety of the patient. Some of the more vital matters from both sets of legislation are discussed below.

Prior Risk Assessment within IRR'99 there is a requirement that 'a new activity involving the therapeutic use of radiation may not begin until a risk assessment has been made and recorded in writing. This assessment must be kept up to date where there is any significant change in equipment or its operational use.' This assessment will normally be drawn up by RPA together with the RPS and include the following:

a) A description of the work,

b) Exposed groups,

c) Hazard assessment, including dose rates,

d) Control measures, such as access to the roof of a linac where the dose rate may be high. Such measures will also be identified in the local rules.

As part of a safety strategy and assessing risk the Royal College of Radiologists has produced: '*Towards Safer Radiotherapy*' (2008). Some recommendations within this document are:

◆ All departments should have an externally accredited quality management system;

◆ Good multidisciplinary working with clear communication is essential;

- ◆ The introduction of new techniques (see below) needs to be carefully planned;
- ◆ Each department should have a system for reporting and analysing errors.

The document proposes a new system within the UK for reporting and analysing errors.

Many of these recommendations are also within regulations and cancer standards.

16.12.1.1 Critical examination (IRR'99) and quality control baselines

Before new equipment is used clinically a critical examination must be performed and documented which ensures that the equipment is working safely both for staff and patient. Associated with this is the commissioning of the equipment for clinical use and also the testing of the equipment that will form the QC baseline for the QA programme (see Chapter 11).

16.12.1.2 IRMER 2000 (see also Chapter 15)

This legislation has been written to ensure that all patient exposures are justified and optimized. Various roles are identified within the legislation. In particular this legislation stresses the need for a set of procedures. These include the identification of all staff, their roles and their training.

The radiotherapy department should describe and list the areas of competence across all relevant treatment modalities into which their processes will be divided for the purpose of a competency based training system. For each area of competence, it should list any objective criteria which need to be fulfilled for awarding the documentation of competency in that area besides just the opinion of the authorized assessor.

16.12.1.3 New technology (mentioned in both sets of legislation)

The above guidance is vital when bringing in new technology or changing techniques on equipment already in use. The quality in particular, will ensure that appropriate protocols and procedures are in place to ensure safe working practice before any new equipment or techniques are used clinically. The introduction of IMRT is a good example. Techniques using IMRT have been in use for 10–15 years in the USA. However, a recent audit in the USA has identified shortcomings in the accuracy of dose delivery and highlights the shortage of physicists. In the UK, centres have generally taken longer to commission IMRT. There is now a new initiative to ensure that UK centres can attain a high standard of excellence in IMRT planning and delivery by having appropriate training and audit.

Generally new technology requires:

- ◆ The collaborative efforts of an multi-disciplinary team,
- ◆ Carefully designed procedures and protocols,
- ◆ Appropriate training,
- ◆ Following through the first few patients, possibly with an external expert,
- ◆ Audit.

16.12.2 **Accuracy generally**

It is now generally accepted that the levels of accuracy required for radiotherapy are as follows:

- Dosimetric accuracy: 3% (1 standard deviation (SD)) is the currently recommended requirement on the value of the dose delivered to the dose specification point (ICRU prescription point). For the dose at other points a figure of 5% (1 SD) can be taken.
- Geometric accuracy: 4 mm (1 SD) and 5mm (1 SD) have been discussed as practical limits in the literature.

16.12.2.1 **How is accuracy maintained?**

As far as dosimetry is concerned this is mainly the responsibility of the physicists and dosimetrists. The various aspects of this are:

- Commissioning of all equipment and the links between equipment, e.g. CT scanners to linacs; treatment planning to linacs;
- Establishing routine quality control measures;
- Working to national Codes of Practice for dosimetry
- Maintaining dosemeters and links to secondary standards (see Chapter 7).

Regarding treatment planning, dosimetry and geometry, all staff need to understand where minor errors can occur and have rules and checking procedures to minimize these errors. Similarly for the delivery of radiotherapy, the treatment staff need to understand the importance of maintaining geometric accuracy to within 1–2mm and the role of IGRT in this.

However, whatever steps are used to maintain accuracy there is still the chance of an error, human or otherwise, that will affect the outcome of treatment both tumour control and critical structure morbidity.

16.12.2.2 **Radiotherapy incidents and errors and error reporting**

IRR99 and IRMER 2000 require that radiotherapy departments report when incidents or errors occur. Each department will have its own internal reporting arrangements to its NHS Trust. But if the error results in an overdose there is the need for mandatory reporting. The criteria for this type of report is, at present, stated in PM77 which is written with Reg 33 (IRR'99) particularly in mind—for errors that occur due to malfunctioning equipment—but also is used to define when an error must be reported under IRMER 2000. PM77 defines overdose as overexposure of 10% for a complete course of treatment and 20% for any single fraction that has been overexposed. These refer to the *actual treatment exposures* not the imaging doses within the radiotherapy procedures for which there is not yet any guidance. Only overexposures have been identified in PM77, even though one of the major incidents within UK resulted in underexposure of many patients with potential for subsequent lack of tumour control. Attempts have been made to improve the criteria set within PM77 by including factors such as under dose, the overexposure of critical structures, geometric misses etc, but to date no such document has been formally accepted by the regulatory bodies. However, individual centres continue to report all incidents to their own hospital trusts and voluntarily to data bases such as RCR. This is to allow for learning by other centres and to keep records for the whole of the UK. This requirement has now been included in the Peer Review Standards for Radiotherapy, i.e. there should be a post incident procedure for the department which specifies at least the following:

♦ That it should follow up a Level 1or 2 incident (as defined in *'Towards Safer Radiotherapy, Figure 3.1, p. 20–21*).

♦ The personnel responsible for carrying out the post incident procedure.

♦ That a root cause analysis of the incident is undertaken (as defined in: Taylor-Adams S, Vincent C. (2004). Systems Analysis of Clinical Incidents. London: Imperial College Clinical Safety Research Unit).

16.12.3 Audit/inspection

Audit, both internal and external, plays a vital role in a quality system and as a general check on the radiotherapy processes within a department. It ensures a fuller understanding of the various processes and provides a check on certain outcomes to give assurance that procedures are working safely (and also within the law). Various documents make the requirement for audit, but particularly within IRMER 2000, which requires duty holders to take part in a clinical audit programme. It also places responsibility on all staff involved in a medical exposure, including those acting as practitioner, to participate in a programme of audit to demonstrate compliance with established protocols and procedures. Examples include in-vivo dosimetry results, checks on process (in particular checking on signatures and document control), clinical governance audits.

16.12.3.1 What should be reviewed in a quality audit visit?

The content of a quality audit visit should be predefined and will depend on the purpose of the visit; for example, is it a routine regular visit within a national or regional quality audit network, is it regulatory or cooperative between peer professionals, is it a visit following a possible misadministration, or is it a visit following an observed higher than expected deviation in a mailed TLD audit programme that the centre cannot explain?

For example, a quality audit visit may contain:

♦ Check documentation, e.g. the contents of policies and procedures, quality assurance programme structure and management, patient dosimetry procedures, simulation procedures, patient positioning, immobilization and treatment delivery procedures, equipment acceptance and commissioning records, dosimetry system records, machine and treatment planning data, quality control programme content, tolerances and frequencies, quality control and quality assurance records of results and actions, preventive maintenance programme records and actions and patient data records, follow-up and outcome analysis.

♦ Check infrastructure, e.g. equipment, personnel, patient load, existence of policies and procedures, quality assurance programme in place, quality improvement programme in place, radiation protection programme in place and data and records.

- ◆ Carry out check measurements of beam calibration, field size dependence, electron cone factors, depth dose, electron gap corrections, wedge transmission (with field size), tray factors, mechanical characteristics, patient dosimetry, dosimetry equipment and temperature and pressure measurement comparison.
- ◆ Carry out check measurements on other equipment, such as the simulator and CT scanner.
- ◆ Assess treatment planning data and procedures. Measure some planned distributions in phantoms.

16.12.3.2 Quality assurance in clinical trials

Clinical trials often require new techniques to be used. When the trial is multi-centre it is vital to have a quality assurance programme to ensure accuracy and consistency of treatment planning and delivery across all sites. QA for UK radiotherapy trials is detailed at www.rttrialsqa.dnsalias.org. The main levels of QA are:

1. Baseline questionnaire showing the facilities, staffing and equipment available within the individual radiotherapy centre.

2. Specific trial questionnaire, which may also include the request to develop a process document for each centre to identify how it will plan and deliver the required radiotherapy.

3. Use of dummy runs which require the individual centre to: a) outline various volumes on CT (and maybe MR in the future), including critical structures; b) produce a dose plan that fits the various dose volume requirements.

4. Use of site visits which will: a) audit the process; b) measure reference dose levels; c) measure dose at points in an anthropomorphic phantom and compare these with calculated doses.

Appendix A—Accuracy and Precision

Dr Niall MacDougall

Accuracy and precision are two words which are often used interchangeably in day to day speech. However, in scientific terms each has a specific and independent meaning.

A measurement can be highly accurate and not be precise or any combination of high/low accuracy and high/low precision, which may not make sense at first.

Figure AP1 shows the four possible combinations for accuracy and precision. In this case the example is of 12 arrows fired at a target, aiming to hit the triangle in the middle. Precision is a measure of how close the arrows are grouped on the target. Accuracy is how close the whole group of arrows is to the centre of the target.

From the arrow and target analogy we can also see that it is not possible to ascertain accuracy from one arrow, unless one has knowledge of the precision of the person firing the arrows.

All measurement techniques require their accuracy and precision quantifying before one can rely on individual measurements. This is the case for many examples in this book.

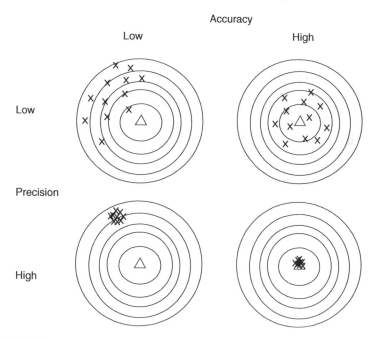

Fig. AP1 High/low accuracy and precision.

Further reading

Professional Guidance

International Commission on Radiation Units and Measurements, *Prescribing, Recording, and Reporting Photon Beam Therapy*, ICRU Report 50. Bethesda, MD: ICRU, 1993.

International Commission on Radiation Units and Measurements, *Prescribing, Recording, and Reporting Photon Beam Therapy, (Supplement to ICRU Report 50)*, ICRU Report 62. Bethesda, MD: ICRU, 1999.

International Commission on Radiation Units and Measurements, *Prescribing, Recording, and Reporting Electron Beam Therapy*, ICRU Report 71. Bethesda, MD: ICRU, 2004.

International Commission on Radiation Units and Measurements, *Prescribing, Recording, and Reporting Intensity-Modulated Photon-Beam Therapy (IMRT)*, ICRU Report 83. Bethesda, MD: ICRU, 2010.

The Royal College of Radiologists, *Towards Safer Radiotherapy*. BFCO(08)1. London, UK: Royal College of Radiologists, 2008. Available from: http://www.rcr.ac.uk/publications.aspx?PageID=149&PublicationID=281.

The Royal College of Radiologists, *A Guide to Understanding the Implications of the Ionising Radiation (Medical Exposure) Regulations in Radiotherapy*. BFCO(08)3. London, UK: Royal College of Radiologists, 2008. Available from: http://www.rcr.ac.uk/publications.aspx?PageID=149&PublicationID=289.

The Royal College of Radiologists, *On target: ensuring geometric accuracy in radiotherapy*. BFCO(08)5. London, UK: Royal College of Radiologists, 2008. Available from: http://www.rcr.ac.uk/publications.aspx?PageID=149&PublicationID=292.

European Society for Radiotherapy & Oncology (ESTRO), Brachytherapy Guidelines. A number of useful guidelines and recommendations on brachytherapy are available from: http://www.estro.org/estroactivities/Pages/Guidelines.aspx.

Statutory Regulations

The Ionising Radiation Regulations (IRR) 1999 (Statutory Instrument 1999, No. 3232). London, UK: HMSO, 1999. Available from: http://www.legislation.gov.uk/uksi/1999/3232/contents/made.

The Ionising Radiation (Medical Exposure) Regulations (IRMER) 2000 (Statutory Instrument 2000 No. 1059). London, UK: HMSO, 2000. Available from: http://www.legislation.gov.uk/uksi/2000/1059/contents/made.

Medicines (Administration of Radioactive Substances) Regulations 1978 (MARSR). Available from: http://www.dh.gov.uk/ab/Archive/MARSR/DH_095135.

Physics Texts

Handbook of Radiotherapy Physics: Theory and Practice, edited by P. Mayles, A. Nahum and J.C. Rosenwald. Published by Taylor & Francis in 2007.

The Physics of Radiation Therapy (4th Edition), by Faiz M. Khan. Published by Lippincott Williams & Wilkins in 2009.

Other 'Radiotherapy in Practice' titles

Radiotherapy in Practice – External Beam Therapy, edited by Peter Hoskin. Published by Oxford University Press, Oxford in 2006.

Radiotherapy in Practice – Radioisotope Therapy, edited by Peter Hoskin. Published by Oxford University Press, Oxford in 2007.

Radiotherapy in Practice – Imaging, edited by Peter Hoskin and Vicky Goh. Published by Oxford University Press, Oxford in 2010.

Radiotherapy in Practice – Brachytherapy (2nd Edition), edited by Peter Hoskin and Catherine Coyle. Published by Oxford University Press, Oxford in 2011.

Index

Printed and bound by CPI Group (UK) Ltd, Croydon, CR0 4YY